MARKETING AGRICULTURAL COMMODITIES

Marketing and price risk management in agricultural commodities

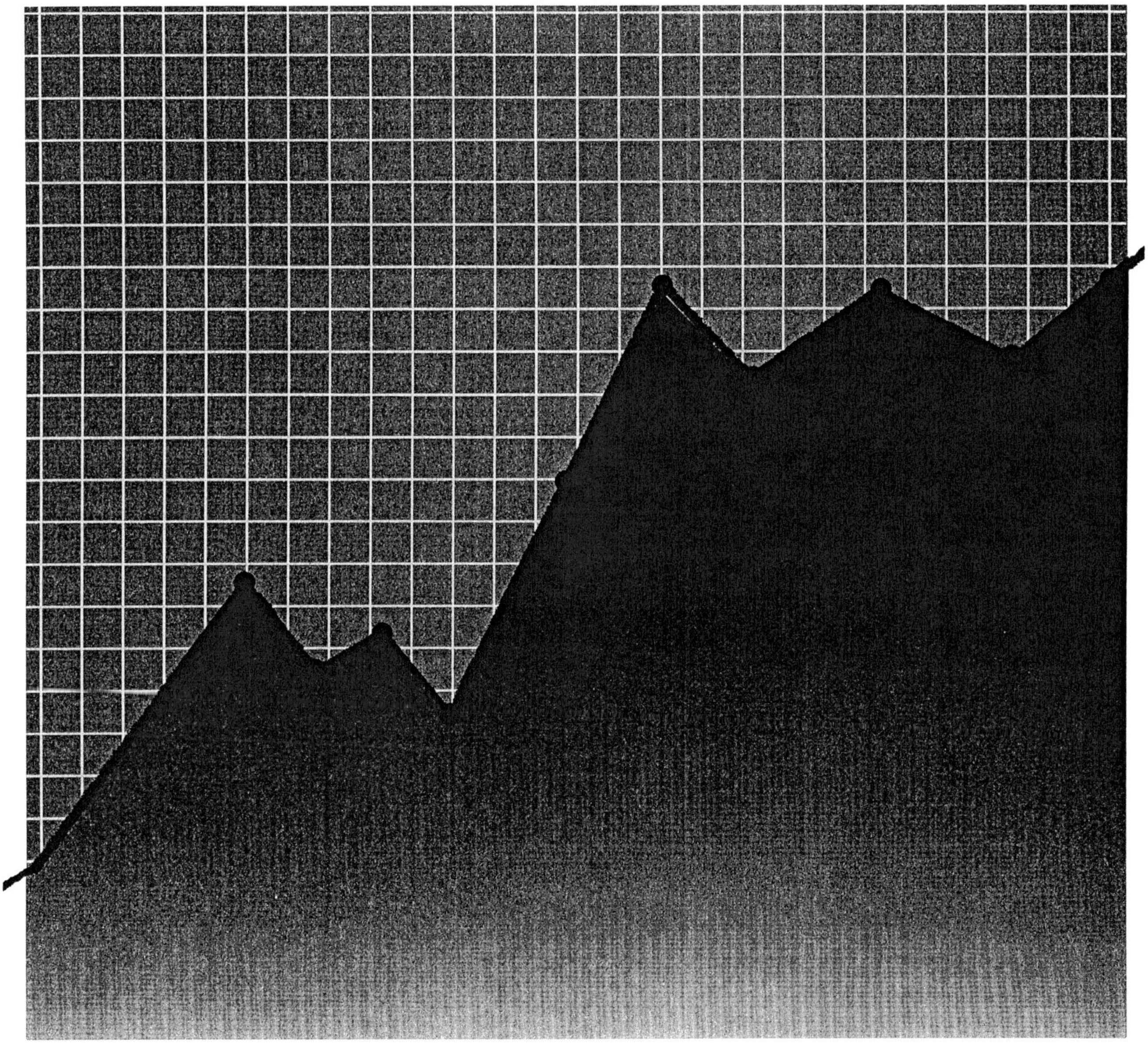

FARM BUSINESS MANAGEMENT

PUBLISHER

Farm Business Management (FBM) is a series of manuals created by Deere & Company. Each book in the series was conceived, researched, outlined, edited, and published by John Deere Publishing. Authors were selected to provide a basic technical manuscript which could be edited and written by staff editors.

PUBLISHER: John Deere Publishing, Dept. FOS/FMO. John Deere Road, Moline, Illinois 61265-8098; DEPT. MANAGER: Alton E. Miller.

EDITORIAL STAFF

Managing Editor:	Louis R. Hathaway
Editor:	John E. Kuhar
Publisher:	Lori J. Lees
Promotions:	Cindy S. Calloway
Art Director:	Bernard F. DeKazel

AUTHOR: *Wayne D. Purcell*, Ph.D., is Professor of Agricultural and Applied Economics at Virginia Tech. He also serves as the Director for Research Institute on Livestock Pricing and Coordinator for Virginia's Rural Economic Analysis Program. He has received eleven teaching awards from 1968-1991, has authored a marketing text, a text on commodity futures and options and over 200 additional publications.

CONSULTING EDITOR: *H. Evan Drummond*, Ph.D., Professor of Agricultural Economics at the University of Florida, has nearly 30 years of professional experience in the U.S. and Latin America. He is best known for his involvement in undergraduate teaching and curriculum development. He is the co-author of an introductory text and recipient of numerous teaching awards. Recent accomplishments include the conversion of an introductory course to a multi-media delivery system.

COPY EDITOR: *Ralph Reynolds*, a freelance writer and photojournalist, was executive editor of *The Furrow*, until his retirement from Deere & Company in 1985. He earned degrees in both agricultural education (teaching vocational agriculture for four years) and journalism before joining Deere in 1959. He is the author of an autobiographic book, *Growing Up Cowboy*, published in 1991 (Fulcrum).

SPECIAL ACKNOWLEDGEMENTS: The author and editor wish to thank Karen O'Conner, Virginia Tech., for her tireless assistance.

CONTRIBUTORS: The Publisher is grateful to the following groups who helped by providing photographs or information used in this text: Globalink, an electronic news service of Professional Farmers of America, Cedar Falls, IA and the United States Department of Agriculture.

FOR MORE INFORMATION
This text is part of a complete series of texts and visuals on agricultural management entitled Farm Business management (FBM). An instructor's kit is also available for each subject. For information, request a free Catalog of Educational Materials. Call 1-800-522-7448 or send your request to: Midland Press Corporation, John Deere Dept. 800, 5440 Corporate Park Drive, Davenport, IA 52807.

ISBN 0-86691-217-7

CONTENTS

8 ANALYZING THE MARKETS

9 A CHANGING MARKETPLACE

10 FINAL THOUGHTS AND A BIT OF MOTIVATION

APPENDIX

PREFACE

This book is written for the current and aspiring American farmer. The concept of **marketing** can be big and complex. Reference books define marketing as all of the economic activities that go on to get the product from the point of original production to the final consumer. There is, thus, an entire system of marketing that starts with the production activity of the farm on the bottom end and finishes with a food or fiber product ready for the consumer's use on the top end. Entire books have been written about marketing systems and what we need for them to be efficient. There is coverage of pricing efficiency, price discovery, and pricing to value. These concepts may get defined later, but the key point here is that it can get complicated.

It is important to understand that the farmer is the first link in the chain of events that becomes a marketing system. The final link is the consumer. It will be necessary to understand some of the characteristics of this system. In particular, it is important to understand that the producer is always facing a price at which he can sell his commodity, and that price comes out of and is generated by the marketing system. Thus, it is important to have some appreciation of **price**, how it is determined, and why price is at the level that is being observed at any particular point in time. Having said all that, the focus here is on the marketing decisions of the farm producer.

- **What does it take for that producer to make effective marketing decisions?**
- **How important is it?**
- **Is this the way that farmers can differentiate themselves from their neighbor down the road, make a decent income, and be viable when other farmers are struggling?**
- **How much potential is there if the farmer as a decision maker starts to get as good in making marketing decisions as in making increasingly sophisticated production decisions?**

The objective of this book is to move the individual farmer as a decision maker to a substantially higher ground in terms of marketing know-how and marketing effectiveness. This is based on the assumption that there is more potential for that individual farmer to improve his economic position by improving marketing than in still further improvements in production. Most farmers who are in the business full-time are very capable in terms of production know-how and technology. They use sophisticated chemicals, and growth promotants. They understand the environmental consequences of various types of application techniques. They are not only good and efficient producers but are good stewards of the land. Somehow, it seems that know-how on the production side is relatively easy to get. It is available from the local extension agent who historically has been trained primarily in agronomy or animal science or in some of the other production-oriented education programs at the state university. It is available via written materials from the company that breeds the seed corn or the company that makes the fertilizer or the herbicides or the insecticides. Those private firms employ very knowledgeable scientists to write about how to use their products effectively in the production process.

Most producers are also fairly well informed on financial management. Banks and other credit institutions are increasingly working with the farmer to prepare a cash flow analysis. Many extension agents know financial planning procedures and can help integrate the production plan and the cash flow projections in a financial plan.

If you switch to the marketing side, however, far less insight and expertise is available to the individual farmer. They are getting better, but many of the local agricultural extension agents still have little or no training or experience in marketing and marketing strategies. The seed company that produces the hybrid seeds is not expert in handling the very risky and volatile markets that the farmer faces when he gets ready to sell his corn, soybeans, or wheat. The scientists employed by the firms that provide equipment and seed and herbicides to the farm sector are understandably trained in production techniques. They are not trained and experienced in marketing and are not necessarily sophisticated observers of the marketplace.

What all this means is that the individual farmer has tremendous potential to improve his or her position by becoming better in marketing. This book is dedicated toward providing the basics and the applications that the individual will need. It will not always be easy, but there is nothing very mystical about markets and marketing. If as much energy is applied to this need as has been applied to staying up-to-date on the production side, the payoffs will be tremendous. As the individual farmer gets better in handling marketing and marketing decisions, American farmers as a group become more competitive. They fare better in the growing competition in a world market. Agricultural products continue to be a shining light in terms of our ability to profitably export to overseas markets. Without question, as we become more nearly a global market and as communication techniques shrink distances, it is going to be important that the individual farmer and farmers in general become better marketers than they have ever been before. The potential is huge. The opportunities are before us. The need is to recognize that potential and those opportunities and to go after them. This book is dedicated to helping you get there.

1

Marketing And Why It Is Important

INTRODUCTION

Technically, marketing is all the economic activities that are required to put the farm product in the hands of the consumer. This involves processing, transporting, financing, storing, and all of the various things that are required to take a raw material at the farm level and turn it into something the consumer wants. While it is useful to understand all of those various activities and the very wide scope of marketing in its entirety, the individual farmer has to back off and appreciate how he interacts with this nebulous thing called "the market." The farmer is a producer first, and he has food or fiber products to sell. Somewhere during that production process, sometimes even before the production process starts, that individual has to think about marketing that product or that commodity. In this beginning chapter, we will think about marketing from the viewpoint of that individual farmer and in terms of the first connection with someone to whom the food or fiber product is sold.

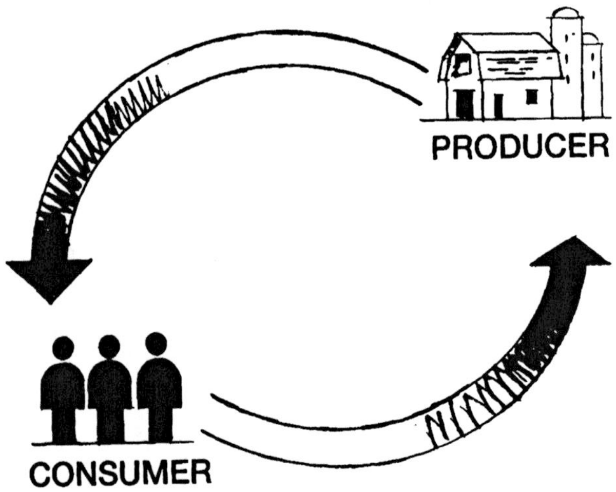

Marketing involves all the activities from producer to consumer

What we are saying is that the farmer needs to be both a producer and a marketer. Producing the product and then just hoping that the price will be good enough to generate a profit when the product is sold is not what we need. To be a marketer and to deal with marketing, the individual has to back off from that action of just selling and **prepare to market the commodity**. Later on in the book, there will be talk about a **marketing plan** and how to lay out a coordinated set of actions so that what the product is sold for will typically be adequate and profitable. Historically, as suggested, the emphasis at the farm level has been overwhelmingly on production. There has not been nearly as much attention paid to marketing. There is a tendency for the individual to throw up his or her hands and say, "It is too much; I can't deal with it." But that is clearly not the case if you get past that initial perception of vagueness and complexity.

Most farmers are already sophisticated producers

MARKETING'S POTENTIAL

Let's look at an example of how much potential there might be to doing something better in marketing as compared to production. A corn farmer, for example, has total cash costs in planting, cultivating, caring for, and harvesting a corn crop of about $175 per acre. This doesn't include all of the fixed costs of the farm, it just deals with those that you have to pay in a particular year because you are planting corn. It covers the seed, fertilizer, labor, upkeep and fuel on the equipment, and any fixed debt payment that you have to pay on an annual basis to the local bank because you are a farmer. All this runs about $175 an acre. If this farmer gets 150 bushels on average per acre, simple division ($175 ÷ 150) suggests that those cash costs are running about $1.17 per bushel. If the farmer could get his yields up to 200 bushels (and that is a huge jump that would be very difficult to accomplish, and would probably incur substantial additional costs on a per-acre basis) costs change dramatically.

It's hard to get yields that high

Let's assume the unlikely would occur and the farmer is able to become a better producer and get yields up to 200 bushels per acre and the per-acre costs are still $175. If we divide $175 by 200 bushels, we get $.87 per bushel in terms of cash costs. If we subtract $.87 from the earlier $1.17, we find that the farmer's position has improved by $.30 per bushel. *To get that improvement requires a huge and very unlikely increase in production efficiency and production effectiveness.*

Contrast this, for example, to the situation that is normally there every year for most corn farmers. Opportunities to price the corn crop will vary anywhere from $.50 to $1.00 per bushel--sometimes much more--in almost every production year. It is very unusual to find a year in which the prices offered to the individual farmer at the local elevator don't move as much as $.50 per bushel from, say, February and March, when you are thinking about planting the corn, all the way until harvest time in October. *What this suggests is that opportunities for improvement on a per-bushel, and therefore on a per-acre, basis in marketing are typically far in excess of what might be possible in terms of improvement from becoming a still better producer.*

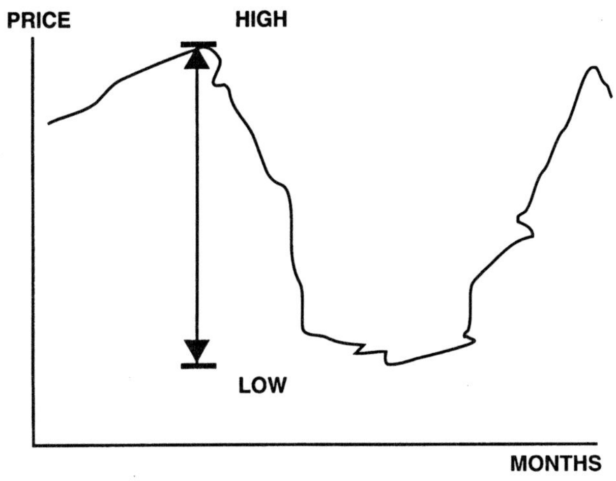

Graph 1 - Prices offered during the year always vary a great deal, creating opportunities

THE BOTTOM LINE

The message is clear. For the individual farmer who wants to become more competitive, become more viable in an economic context, and to survive and succeed in farming, you have to look at getting better in making marketing decisions. As the going gets tougher and it is harder to make a profit it will be marketing skills that can bail you out and let you keep that farm.

The same thing is true in the livestock commodities. It is not unusual, for example, to find per-head profits on cattle feeding programs that vary by as much as $100 per head within the year. Basically, we are talking about an 1100-1200 lb. steer being sold. If the range within the year from profits to losses is, for example, $88 per head, we are talking about a differential of about $8.00 per hundredweight for an 1100 lb. steer. Think about the potential to do something about that variability in profits. Most of the opportunity comes in doing better in marketing as compared to doing a still better job of production.

The feed costs of gain in carrying a steer in a feedlot from, say, 700 to 1,100 lbs., runs around $50 per hundredweight. Since there are 400 lbs. of gain, the feed costs during the steer's typical stay in a feedlot would run around $200. This will change, of course, with the cost of corn and with protein supplement, but let's use this figure for purposes of illustration. Above, it was indicated that because of selling prices in the marketplace and the fact that the market is so volatile, the variation in profits in a cattle feeding program within a year can easily be $100 per head. Think, then, about what you would have to do to influence the profits by as much as $100 per head on the production side.

Feed is the big cost, and it is hard to decrease feed costs much

If the feed costs are $200 during the 120-150-day stay in the feedlot, to change the profit picture by $100 per head, you would have to cut the feed costs in half. That means that the per-hundredweight feed cost of gains drops from $50 per hundredweight down to $25 per hundredweight. It is technically impossible to reduce the feed costs of gain that much through improved conversions and improved feeding efficiency. It doesn't matter how good the cattle are or what breed they are, how effective the rationing, and how effective the management is, you simply can't influence cattle feeding profits by $50-100 per head by changing production efficiency. But the marketplace routinely influences profitability by this much within a year because prices are so variable.

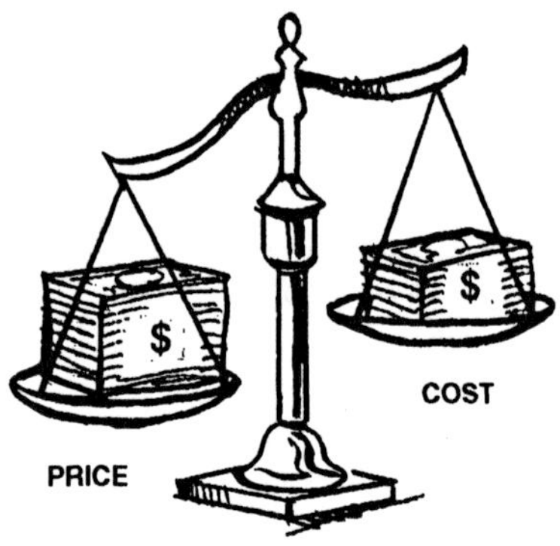

Market price outweighs changing costs in boosting profits

KEY TO SUCCESS

The point is that the big potential lies in marketing. This is not to suggest that being the best producer that you can possibly be is not important. Of course it is. Being competitive in terms of efficiency of production is a necessary condition to being a successful farmer. The *sufficient* condition to being a highly successful farmer is being better than your neighbor and better than average as a marketer. That is what this book is about, and that is why it is so important to deal with this concept of marketing in some detail.

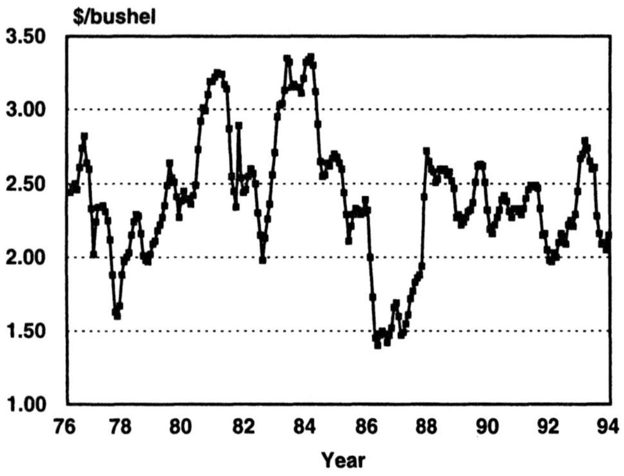

Graph 2 - Monthly average corn prices, 1976 to 1994

Graph 2 shows monthly average corn prices since the mid-1970's. This figure is introduced to provide some real-world evidence of how variable selling prices are in the marketplace. It is useful for the reader to think about how much of the variability might be due to differences in production efficiency. It would be substantial, of course. But across producers, it in no way approaches the potentials associated with doing something about the price variability that the farmer faces. There are a number of periods within this plot during which prices dipped far below cost of production for even the most efficient producer. We will come back later in the book and deal with some of these issues in more detail, but an important point can be made here.

In almost every instance, there was an opportunity for the individual corn farmer to get price established at profitable levels sometime prior to or during the production period.

Don't be a price taker

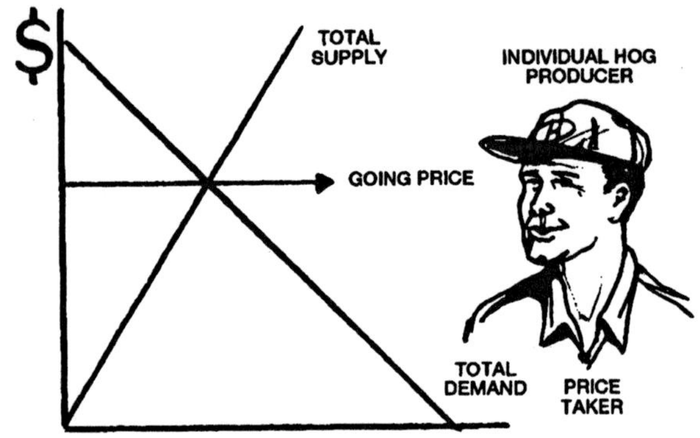

If you wait, you have to take the going price

It is a matter of not just waiting until the day the corn is harvested, hauling it to town, and taking whatever the bid is at the local elevator. Just "waiting" is what creates the big problems for farmers. Then, they are exposed to whatever the marketplace happens to be willing to pay on that particular day, and *the farmer is very much in a price taker status.* Again, it is becoming apparent why getting some expertise in marketing is so important.

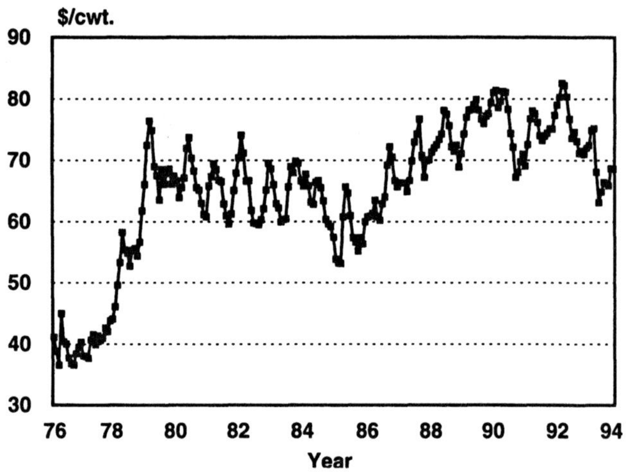

Graph 3 - Monthly average choice slaughter steer prices, 1976-1994

Graph 3 shows monthly average prices for Choice grade slaughter steers since the mid-1970's. This figure is introduced to remind every farmer that variability in the livestock sector is just as great as it is in the grain sector where weather problems often create the price variability. In livestock, part of it is a matter of corn costs "leaking" over into the livestock sector and having an influence on feed costs. Part of it is caused by producers making production decisions and having to bear the price consequences of those decisions, especially when the decisions are to increase production.

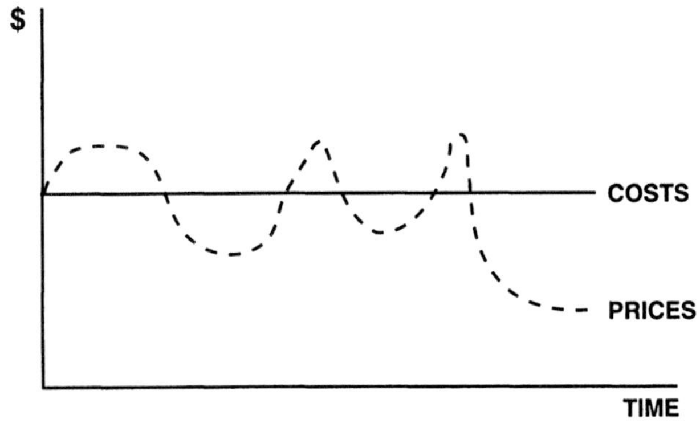

Graph 4 - A price taker often sells at a loss

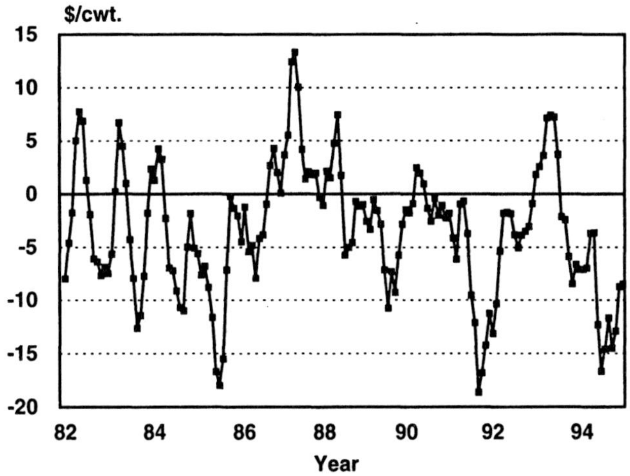

Graph 5 - Profits above costs for typical custom cattle feeding program, 1982 To 1994

The extreme volatility that is shown in the mid-1980's demonstrates how difficult it is for a cattle feeder. Part of the downward moving prices in 1986 were associated with a government buy-out program in the dairy sector that brought a number of slaughter cattle to market unexpectedly. But there are many instances, and 1985 is a good example, in which prices moved dramatically when there was no external influence such as a government program in the dairy sector. The prices are simply very volatile.

The business is at risk

Graph 5 follows the volatility shown in Graph 3's trace of cattle prices with a plot of profits above all costs for a typical Southern Plains custom cattle feeding program since 1980. The USDA publishes these margins in their Livestock, Dairy, and Poultry Situation and Outlook Reports, and the variability facing the cattle feeder is clear. The data in Graph 5 are monthly averages. Individual feeders' losses can approach or exceed $10 per hundredweight, which would add up to $100 per head or more. The risk associated with cattle/feeding is huge! Again, to be successful, the entrepreneur has to be prepared to do something about the very volatile prices.

The emphasis here, then, is on the marketing decision facing the individual farmer or producer. But the importance of understanding the marketing system in a bit more detail and in a somewhat broader context has also been introduced.

The price plots in Graphs 2 and 3 and the highly variable profit plot in Graph 5 ought to raise questions in the minds of any interested reader. Why are prices so variable? What is the source of the variability? Why is it that the individual producer can apparently be exposed to prices that are substantially below their cost? How could this be? There is a general notion that entrepreneurs and business firms respond to profit opportunities. If there are no profits, then why are people involved in farming? And if there are no profits, how is it that people stay in farming? Perhaps most importantly, why is it that the farmer does not consistently get a price that at least covers cost of production? To establish a framework within which to think about these very difficult issues, we have to talk about some very basic economic principles. We have to understand the most fundamental dimensions of demand and supply and how price is discovered within a **demand-supply framework**.

We have to understand why these prices can go so low

DEMAND, SUPPLY, AND PRICE

At any point in time, buyers will take more of a product only at a lower price. This is true whether you are bidding on baseball cards at a local auction, buying antiques, or buying weaned calves to be grazed until the following fall. If, as you approach that particular opportunity to buy, you find that you can buy each unit cheaper than you expected, you are likely to buy more. You take home more of those antiques or a bigger load of those weaned calves.

This is logical and intuitive. *Buyers as users tend to take a bigger quantity at any point in time only at lower prices.* This is true whether you are talking about an individual auction sale as mentioned above or whether you are talking about a much bigger and aggregate level demand for a basic product such as corn. The companies that turn the corn into syrup, turn it into feed, or put it into a bunker to feed cattle or hogs will tend to buy more of that feed only if they can get it at a lower price. Extending that thinking suggests that the situation at the world level is the same. The only way you can send more corn to Mexico, to China, or to any other potential buyer in the world market at any point in time is through a lower price.

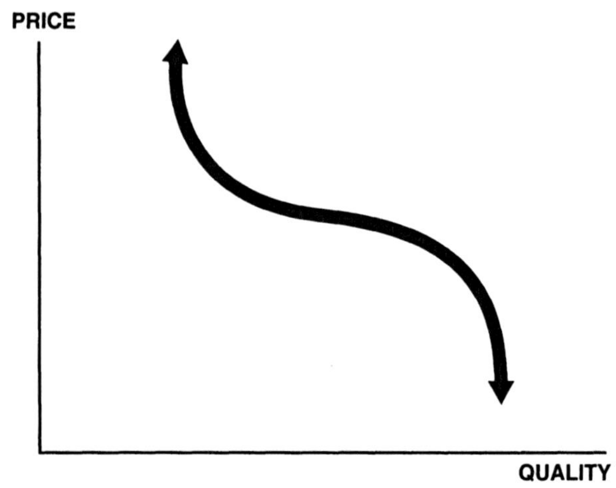

Graph 6 - If you want to sell more quantity, price has to decline

THE LAW OF DEMAND

What this means is that there is a demand for any of our food and fiber products that adheres to what is sometimes called the **law of demand**. **The law of demand states: buyers will take more only at lower prices**. It is sufficient in this particular discussion to recognize that this means **the demand for a product is a schedule of the quantities that will be taken at alternative prices**. This "schedule" is developed with buyers' incomes, prices of other products, and preferences held constant. These are called the *ceteris paribus* conditions in the texts.

Price per Slice	Slices Consumed at Lunch
$2.00	2
1.50	3
1.25	4
1.10	5

Table 1 - The demand schedule for a typical consumer of pizza on his/her lunch break

Table 1 provides an example, using slices of pizza. As prices go down, note that this particular consumer, as would be typical of any buyer, will tend to take more slices. The consumer's income is not changing, he or she has a particular "preference" for pizza, and prices of substitutes (burgers, etc.) for pizza are not allowed to change.

I have only so much money right now

If we think about that, whether it be the demand for corn, for automobiles, for baseball cards, or whatever, we recognize that at the aggregate level, the same characteristics will hold. More corn can be moved through U.S. marketing pipelines in one week compared to the previous week only if prices are lower. This will always be the case. What we have established, hopefully, is that the demand for any particular food and fiber product is, in fact, the entire schedule of quantities that will be taken at alternative prices. If you plot it such as in Graph 7, this demand curve or demand surface slopes down and to the right. It has a negative slope.

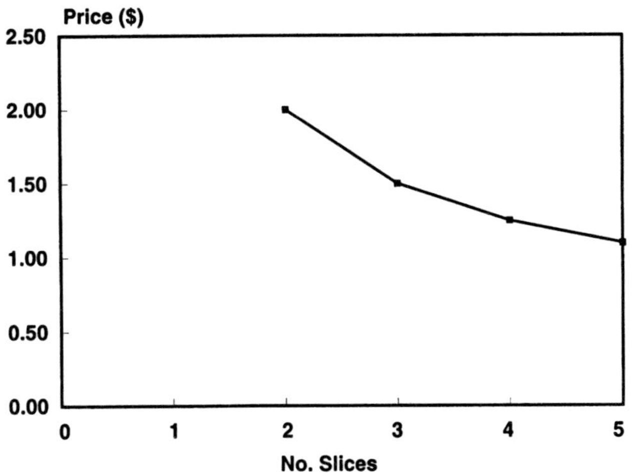

Graph 7 - Demand curve for pizza for a typical consumer

For the time being, that is as deep as we need to get into the discussion of demand. What we have to start to appreciate is that buyers of our food and fiber products will decide the quantities they will take at the prices being presented at this particular point in time.

THE CONCEPT OF SUPPLY

We are now starting to close the loop and starting to understand where price comes from. The quantities that are being offered for sale at any particular point in time at alternative prices, is called **supply**. This "schedule" is for given input costs and for a given level of technology. If these change, we will find later that supply shifts. Here, if we can establish an intuitive and basic understanding of the concept of supply, then we have a very simple but powerful framework to help us understand prices and pricing.

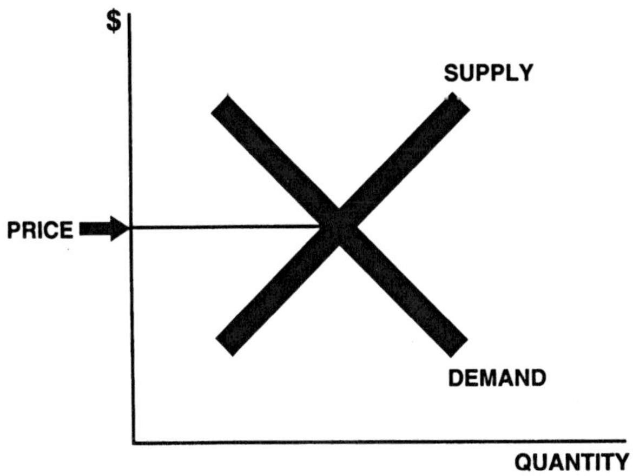

Graph 8 - We get a price when quantity supplied equals quantity demanded

Any producer of any food and fiber product would tend to be willing to produce and offer more at higher prices. This is true for a corn producer, hog producer, vegetable producer, or catfish producer. Any such producer would try to push his production, push his equipment, and push his labor, toward producing more if he expects to be able to sell at a higher price. But as you push a production process, what you find is that it is more and more difficult to get expanded output.

It gets harder as you try to go farther

The process gets less efficient, less and less responsive. You only have so many tractors, equipment, people employed, acres to work with, or so much space on a feeding floor if it's a poultry program or hog program. Thus, as you push, you find that your ability to respond to still further price increases tends to get smaller and smaller. If you translate this into the notion of a supply curve, which as noted above is technically the schedule of the quantities that will be offered at alternative prices, then you have a supply function that looks like the one shown in Graph 9. Note it slopes up and to the right and has some curvature suggesting as you push harder trying to produce and offer more at still higher prices, you have some physical and biological constraints on your ability to respond.

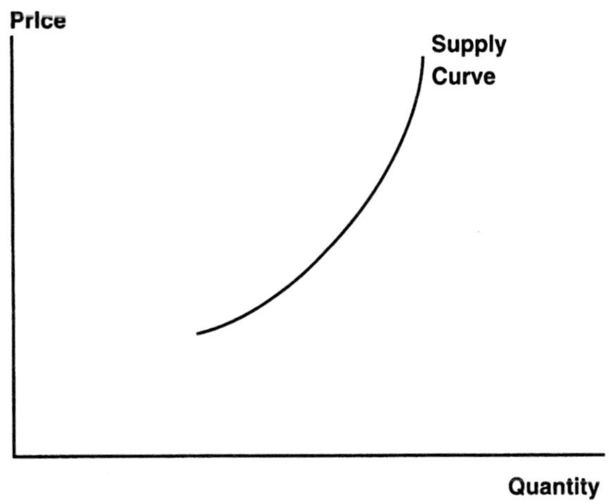

Graph 9 - Typical shape for a supply curve for farm commodities

To review, we have at this point an understanding that

- **Buyers will take more only at lower prices.**
- **Producers will try to offer more if they think the prices are going to be higher.**
- **There are some constraints on how much more producers can produce and offer.**

THE EQUILIBRIUM PRICE

If we take these two curves and superimpose them on each other, we have a supply-demand framework. This process of superimposing the demand curve and the supply curve has been done in Graph 10, and we identify a particular price on the vertical axis. We call it PE, and this is the very important **equilibrium price**. It is the price, the only price, at which the quantity offered by supplies matches the quantity that will be taken by buyers.

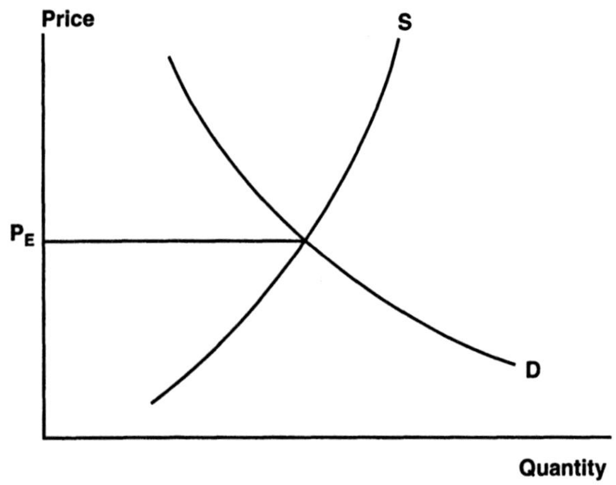

Graph 10 - An equilibrium price when demand and supply are balanced or matched

This is all very basic and very important. What we have is a framework that shows the willingness of buyers to take a particular product. That is called demand. What we also have is a framework that shows producers' willingness and ability to offer varying quantities of product at different prices or expected prices. That is called supply. There is, to repeat for emphasis, only one price at which the quantity that buyers are willing to take away from the marketplace is exactly equal to the quantity that producers as suppliers are willing to bring to the marketplace. That equilibrium price, PE, is sometimes called, quite appropriately, a **market-clearing price**.

At this point, the reader is encouraged to sit back and think for a moment. What do you suppose happens if, for varying reasons, farmers bring more product to the marketplace? What would happen if that curve that we have identified as the supply curve, the curve marked S in Graph 10, should shift to the right? This would suggest that at every possible price that could prevail, the farmers are willing to bring more quantity to market. This could happen if their costs have gone down because of some new variety of seed, some new growth promotant, or some new technology that allows them to increase production at essentially the same use level of various factors of production such as labor, seed, fertilizer, equipment, etc. If the supply curve does, in fact, shift to the right, and this is shown in Graph 11, then the equilibrium price is lower than the price that was established in Graph 10. *What we now have is one of the reasons that prices can and do change over time.*

I have to figure out why price decreases

FARMERS RESPOND TO PRICE

We can also see one of the reasons why the corn prices plotted in Graph 2 have been so variable over time. It is the willingness of farmers to try to produce more corn if they think the price is going to be higher; also weather becomes a factor. Weather influences yields, and we get the occasional droughts or weather disturbances. The higher prices that result can be seen in 1980, in 1983, in 1988, and again in 1993.

Living expenses must be met

Keep in mind that every farmer has an obligation to pay the mortgage payment on the farm, to replace some equipment, to send the children to school, to buy food and clothing, and to provide a place to live. In others words, there is a cash flow need every year. Farmers need so many dollars just to survive with the standard of living that they want to pursue. One of the tendencies when the price of corn or anything else they are producing appears to be dropping is to try to get costs down, to be more efficient. Given enough time, farmers will often bring in more land or equipment and try to produce even more to generate that needed cash flow. This introduction of new capital (land, equipment, etc.) shifts that supply curve to the right and competition among sellers pushes prices down to prices such as PL in Graph 11. Periodically, of course, weather intervenes, droughts come along, and the supply curve is shifted to the extreme left. Prices are then much higher than normal and much higher than would have been expected. It is an up and down world the farmer operates in.

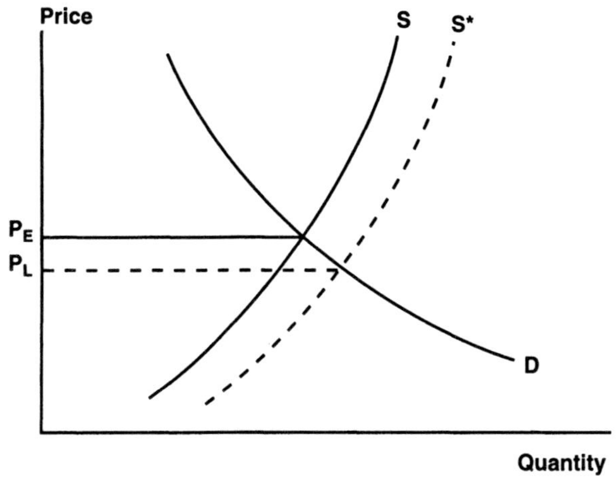

Graph 11 - The price decreasing impact of an increase in supply for corn

THE BIG REASON FOR PRICE CHANGE

Changes in the quantity that is available to the marketplace on a year-to-year basis are the main reason for short-term movements in price of agricultural commodities. It is changes in supply or quantity brought to market that dominate the movements in the corn prices plotted in Graph 2 and the cattle prices plotted in Graph 3. Weather, storms, other things that influence level of production, and farmers' decisions (especially if too many of them decide to expand at the same time and increase production) are among the many things that can and do influence supply. And it is not just weather in the U.S. Droughts in other producing countries, because we are now in a global market, can definitely influence prices in the U.S. Price can move dramatically from year to year.

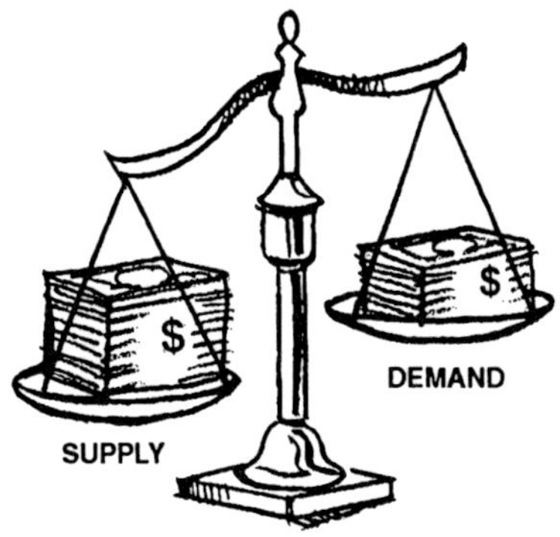

Supply is more important than demand in short run price changes

The other thing that can move price is shift in demand. If you freeze the supply curve and if demand shifts up and to the right, price will go up. Producers would be willing to offer a bigger quantity, and we would have the marketplace at work again to find a new market-clearing price. If demand comes down over time for various reasons, perhaps because the product offering is not adjusted to changing consumer preference then prices tend to go down. If something isn't done to correct the situation, it may be that some producers are forced out of business. We don't have nearly as large a sheep industry, beef industry, or production of some other food and fiber products in the U.S. as was once the case. Often, those *long-run changes* are due to changes in demand as the product for various reasons falls out of favor with a changing consumer either domestically or in the world market.

There is no need at this point to get into great detail. What is important to understand is that in the long term, shifts in demand can and will bring price changes. In the short term, most of the price changes within the year and from year-to-year are brought on by supply changes due to weather or due to producer decisions that move the supply curve along a demand curve that is relatively still and stable. When supply is down, price is up. When supply increases, price goes down. This is the nature of the farming sector and is the source of the high variability that we see in prices of most agricultural commodities over time.

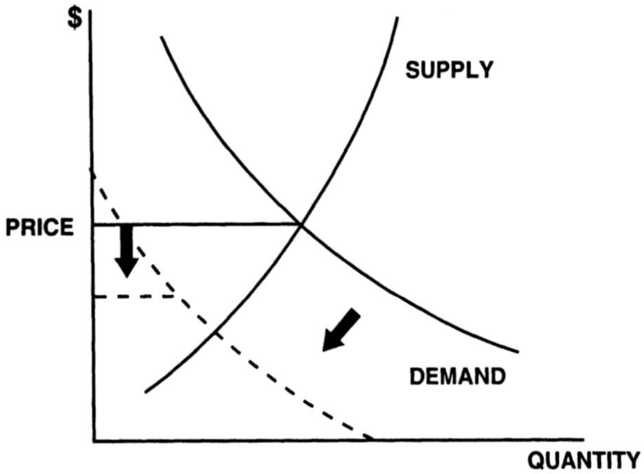

Graph 12 - If supply is constant, changes in demand change price

The importance of all this becomes apparent when you reflect on what supply-demand interactions mean. *Wherever the balance is between demand and supply at any particular point in time, for any particular crop year, for any particular production period in cattle or hogs, that is where the price is going to be when the farmer gets ready to sell his product. What this means, and this is very important, is that the individual farm producer is a price taker.*

FARMERS ARE PRICE TAKERS

They "take" the price that is determined in the bigger, broader marketplace for the corn, cattle, vegetables, or whatever the food and fiber product might be. *Farmers have no power to control price.* They may be very upset that prices of Choice slaughter steers have dipped to $70 per hundredweight, and they need $78 per hundredweight to break even and to cover costs, but there is nothing, as they get ready to sell those cattle, they can do about the price level! You cannot calculate the cost of producing an agricultural commodity and add a margin and then price your product at that level if the price you generate is above this broad market-determined price. You will not sell anything! There are not many other examples in the U.S. economy of an entrepreneur being strictly exposed to the variability and volatility of the marketplace and not being able to do anything about it. But this is exactly the position facing the American farmer. They cannot control price. If, over time, prices continue to be periodically below their cost of production, and they cannot get their costs down enough through increased production efficiency to offset the price pressure, *the marketplace is eventually going to run them out of business.*

The marketplace can be a cruel taskmaster

Graphically, this all looks like Graph 13. This is a two-part diagram that shows an industry-level demand and supply on the right hand side and a price line being extended all the way through to the individual producer. This industry-determined price is the price for which he is going to be able to sell his product at any particular point in time. If this is the middle of October in the middle of the corn harvest, for example, and the price being paid in the cash market for corn in central Illinois is $2.25 per bushel, the corn farmer is going to get something very close to $2.25 per bushel. His price may be a little higher if he is close to a barge loading point where the product can be sent conveniently and cheaply down to export at New Orleans, but he is pretty much saddled with that $2.25 price level. *On that particular day when that product is ready to be hauled to the local grain elevator, this farmer is definitely a price taker. It is a "take it or leave it" situation.*

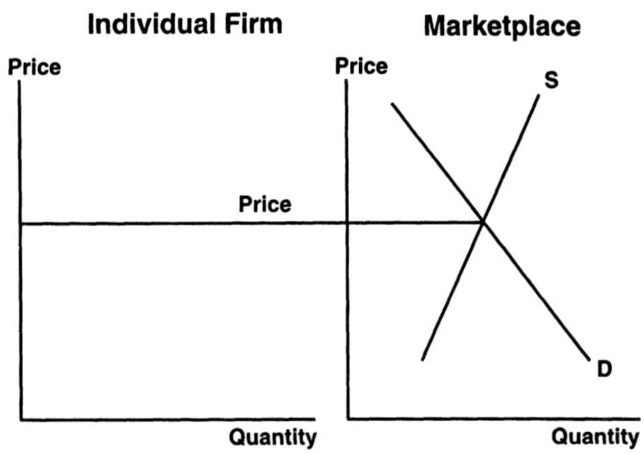

Graph 13 - Demonstration of the price facing the individual producer as it is determined in the marketplace

It is convenient to think about cost levels associated with all this. Graph 14 reproduces the graphics in Graph 13, but it also suggests that an efficient corn producer might have a cash cost of producing corn of $2.00 per bushel because his yields are good and he has fertile central Illinois farmland. If the price is $2.25, this Illinois farmer may be able to survive over time. If you add back all of the fixed costs (and you do eventually have to cover replacements costs for the buildings, machinery and equipment), your costs may go up to $2.40, and the farmer will obviously need, at least occasionally, prices above $2.40 to be able to survive. *But if the cash cost associated with that corn crop is at $2.00, this farmer is going to be fairly competitive.*

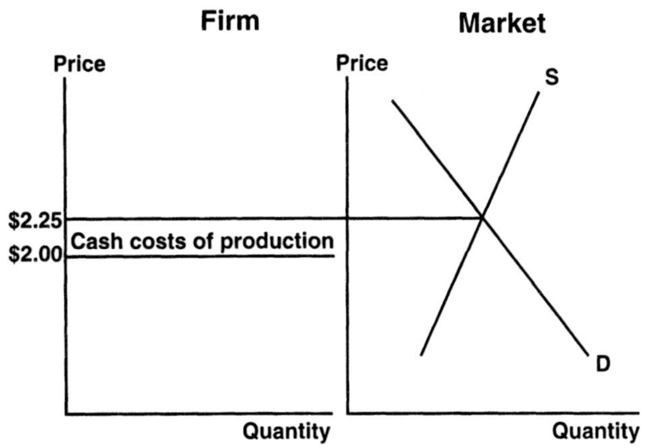

Graph 14 - Demonstration of a low-cost producer who could tolerate $2.25 prices

Contrast that with the corn farmer in the southeastern part of the country, where growing conditions are a bit less favorable, and the cash cost of producing corn is $2.40. If you add back all the other costs, you are talking about a total cost of producing corn at $2.65 or higher. Obviously, having to sell the crop at $2.25 per bushel, or even something slightly higher because of location in the corn-deficit Southeast (where use exceeds production), means that the southeastern corn producer is in trouble over time. But if the only strategy southeastern producers have is to sell the product in the cash market at harvest time, they are in trouble.

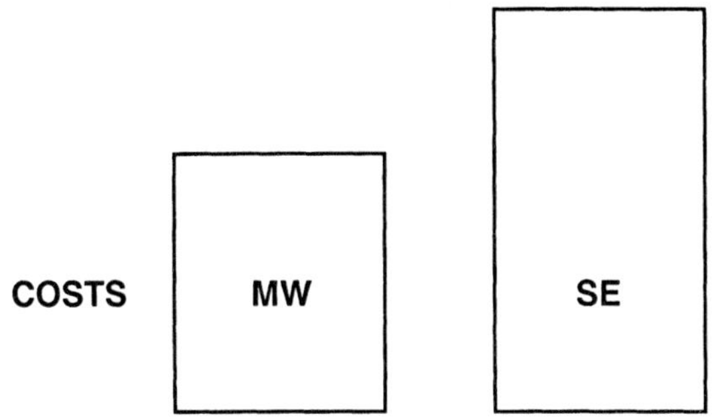

When costs are higher, effective marketing is especially important

CONTROL WHEN YOU PRICE

What is the answer? What can the producer do? The message is that the individual producers have to think about those variables over which they can exercise some control. *If they cannot set price above cost of production, then it is extremely important that they be prepared to manage effectively the one thing they can control: when the product is priced.* This is the essence of price risk management and marketing management for agricultural commodities. It reflects the understanding that if you wait until the last day, all of the flexibility and all the ability to influence anything is gone. What individual producers always have is the right to control when they price. That means being willing to **forward price** and being willing to monitor the opportunities that the marketplace is offering on a day-to-day basis all during the year and then being willing to make a move when an acceptable price is reached. In many respects, this is what marketing management is all about. This need to understand forward pricing and to develop an ability to use forward pricing in an effective price risk management program is the essence of what will be discussed in subsequent chapters.

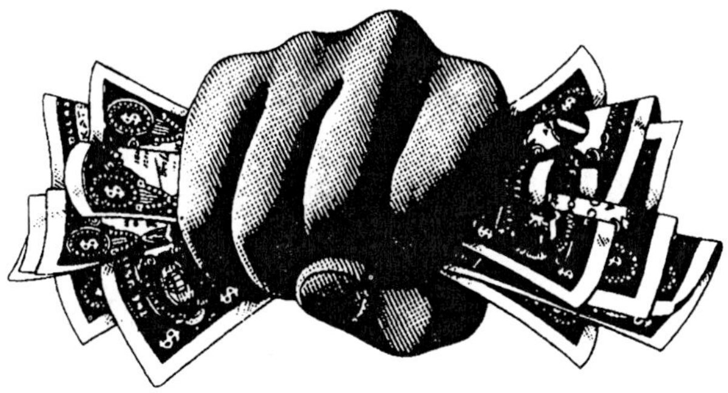

You can keep those dollars by controlling when you price

Before proceeding to those discussions, however, it is useful in this beginning chapter to take a look at some typical marketing decision scenarios. It is not important that every possible alternative that might confront a farmer be covered. It is useful to think about the decisions in some basic food/fiber products including both storable commodities like grains and oilseeds and non-storable commodities like fresh vegetables or the livestock commodities.

SOME TYPICAL MARKETING DECISIONS

Depending on what the product is, the decisions facing the producer and the economic indicators they will have to study will vary. But there are some common needs across all the decisions. Let's look at some situations and then come back and comment in general about some of those common components. Some typical marketing decisions would include:

- **Decision in fresh vegetables on whether to contract with a retail chain that you have sold to in the past or to stay in the open market and try to merchandise your vegetables at a better price when harvest comes;**

- **Decision on whether to sell or store corn as harvest is completed in mid-October and you are looking at what the current cash corn market is offering compared to what might be available if you use your on-farm bins and store the corn;**
- **Decision on whether to forward price soybeans in March before they have even been planted as you contemplate what the marketplace might have in store for you during the coming months and at the following fall harvest; and**
- **Decision on whether hogs should be forward priced as they are moved on to the finishing floor considering what the possibilities might be some weeks later when the hogs will be ready for sale.**

WATCH WHAT OTHERS ARE DOING

This is certainly not a complete list, but it is an example of marketing decisions that the typical farmer faces. There are a number of common components. Without exception, there has to be some consideration given to what you expect producers around the country to be doing. *This is very important.* In the vegetable decision, for example, if you are producing lettuce and you have some reason to think that producers around the country are going to expand lettuce production this year, that might argue in favor of taking an acceptable price through your contacts at the retail chain rather than risking what you might get later in the open market. You understand that if the lettuce supply curve is shifted to the right, prices are going to go down during the year. You have every reason to be concerned if that is the likely scenario, and getting a contract price set is probably the right approach.

Remember to think about what everyone else is planning

The same thing can happen in decisions like storing corn. If a large number of producers decide to store corn rather than sell at harvest, what this will mean is that corn is being shifted out of the October period at harvest into the following May. Supply of corn for the following May is shifting to the right. This can move the price for corn in May down and may generate a situation in which prices do not go up enough from harvest to the following May to cover storage costs.

In the decision on soybeans, regarding what other producers are doing is very important. If there is a big expansion in soybean acreage because prices are starting to look favorable prior to planting, that is a sure way to shift the supply curve to the right and down that sloping demand curve and drive prices lower. Therefore, the prudent decision maker may well decide to protect himself against the influence of many other individual producers making a decision to expand. Get those soybeans forward priced.

Too many acres will bring lower prices

Much the same thing is true of the decision on hogs. It all depends on how many hogs are coming to market 10-12 weeks into the future when these particular hogs will be ready for slaughter.

We have, then, some clear commonality across many marketing decisions:

- **You must be aware of what other producers are likely to be doing.**
- **You must take into account past performance in your own program.**
- **You will need money to repay debt and cover living expenses during this particular year.**
- **You will need to determine your financial ability to carry risk and your related ability to manage intelligently that exposure to risk.**

Think about being different from the crowd

Watching, even anticipating, the behavior and reactions of other producers is important. If corn prices look better than soybean prices, for example, you may want to shift acreage to corn. But every other producer is facing the same situation. What if lots of producers shift, corn prices are forced down, and you get hurt by the collective actions of many producers? You need to step back and recognize what could happen before you shift that acreage.

LEARN FROM HISTORY

Learning from your history is important. A few years back, you were in the middle of expansion in your swine herd. You remember that you were concerned that too much expansion by too many producers could drive prices down--and they did plunge. But you didn't get your hogs forward priced. You don't want to make that mistake again.

Debt and living expenses are critical. If you can't cover those, you can't stay in farming. Any strategy that helps guarantee you will be viable--will be able to pay living expenses and service the debt--needs consideration.

Some managers can't tolerate risk, others seem to thrive on it. Know your own capacities in this area, and remember to keep your financial picture in mind. If you cannot afford to carry the risk of lower prices because you are "stretched" financially with lots of debt, then look for strategies that protect you against that risk.

Keep your attitudes toward risk in mind

If you put the four together, we start to recognize that no one marketing or management program will fit the needs of every farmer. There are some common issues, but every decision maker will be different and unique in some important ways.

Managers need to get more involved in marketing

SUMMARY

In summarizing this beginning chapter, the primary points revolve around (1) the importance of marketing, (2) the difficulties facing the marketing decision given that the individual producer is a price taker, and (3) the need to recognize that good information will be needed in making the marketing decisions. A basic theme of this book is that the potential on the marketing side of the equation to improve the well-being of the operation may often exceed the potential from being a still better producer. If that presumption is accepted, then the immediate need for individuals is to recognize that they cannot control price. Prices of farm-produced food and fiber commodities tend to be very variable. There is no guarantee that the selling price at harvest for any commodity will be at or above the cost of production. *The individual producer is a price taker and has to accept at any point in time the price that is being generated in the bigger, broader market for that commodity. This point is critical, and the reader should reflect on it before proceeding.*

The important message from that recognition is that the one thing the producer can control is when price is established. We will see in later chapters that this will not necessarily guarantee that the program will always be profitable, but it does put the producer back at least partly in control of his or her own destiny. It opens up the possibility of understanding why prices move over time, monitoring those movements, taking advantage of that knowledge and that awareness, and then making a decision as to when price will be established. *Almost anything will be superior to waiting until the last day when the grain needs to be hauled to town, the hogs are at a slaughter weight, or the vegetables have to be picked, and then start to worry about negotiating a price* . There is no flexibility left, no opportunity left, and you have to take whatever price the marketplace is offering on that particular day. *The astute producer can do better than just take what the market offers. To succeed and prosper, an individual must recognize this basic truth and do something about it.*

It is from this perspective and this understanding that the following chapters are developed to move the individual farmer toward a status of being an effective marketer. There are never any guarantees. There is never any one perfect strategy. What will be required is some understanding of how prices move over time and a related willingness to do something about it while there still is some flexibility. The chapters will be developed in that context and in that order. The initial themes will be to sharpen understanding of why prices move over time and provide a framework for the decision maker to use and allow him to take advantage of his knowledge of price movements. Ways of forward pricing will have to be considered, and that means cash contracts, futures markets, or options on futures markets as potentially important parts of a price risk management strategy. By the last chapter, the objective is to have interested readers in a position to be able to compete in highly variable markets, to do a better job than the typical producer, and, therefore, to ensure their viability and success over time.

2

Price Movement Over Time

INTRODUCTION

In Chapter 1, some plots of prices over time were presented. There was also discussion of the basic supply-demand framework within which prices are generated. In a marketplace, it is interaction between buyers, those who want the product, and sellers, those who can offer the product, that determines price. The earlier discussion was at a very basic level, but it was extended to the idea that when prices change over time, there has to be a reason. In the short run, it is usually shifts in supply along a largely stationary demand surface that changes prices. Demand usually does not change very much within the year or even from year to year. In the longer run, there can also be significant shifts in demand that change prices. Buyers' preference patterns change over time and their willingness to pay various prices changes accordingly. The important point to recognize is that price does come from interaction of demand and supply. When prices move over time, it is because demand, supply, or both have changed. One or both of the curves have shifted.

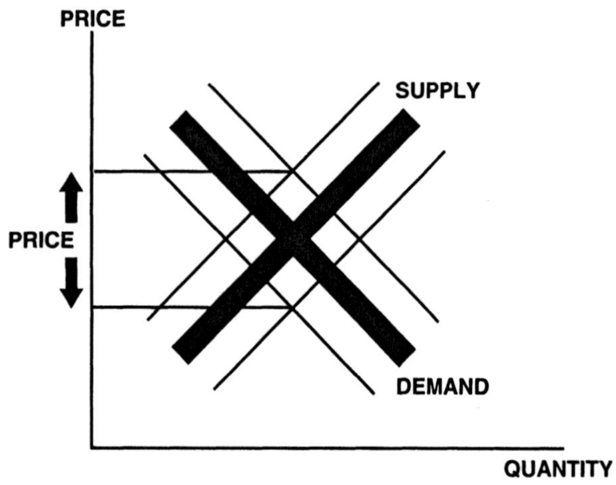

Graph 1 - Shifts in demand and supply bring changes in price

Even within that broad framework, the pricing of agricultural commodities is certainly not an exact science. Consider a situation in which it is early April and a soybean farmer is trying to decide whether or not he should forward price soybeans that he hasn't yet planted. He can call the local grain elevator and get a quote for both soybeans delivered to the elevator that day and a quote for soybeans to be delivered some seven to eight months later in the first half of October. Where does that latter price come from?

Futures contract allows producer to price now and deliver later

FUTURES TRADING MAKES PRICE DISCOVERY POSSIBLE

As we will find in more detail later, the grain elevator is able to offer a price for later delivery because prices for later delivery are being discovered in the futures market at the Chicago Board of Trade (CBOT). On April 10 (or whatever day we might be dealing with), there are buyers and sellers of contracts for later delivery that bring their expectations as to what the October price will be to the trading pits at the CBOT. If they think the price early that morning is too low, they will tend to buy those futures contracts. If they think the price early in the trading day is too high, they will tend to sell those futures contracts. Thus, we have a direct interaction between buyers and sellers in a competitive bid and offer process. Buyers and sellers are actually physically with each other and observing hand signals to indicate a willingness to sell or to buy at certain prices. But it is early April, and no one knows with certainty what the soybean prices will be in early October.

This **price discovery** activity is a dynamic process. Buyers and sellers bring their information on what prices will be in October to the futures exchange, but that information can and will change as time passes. Building of a price based on current information is price discovery.

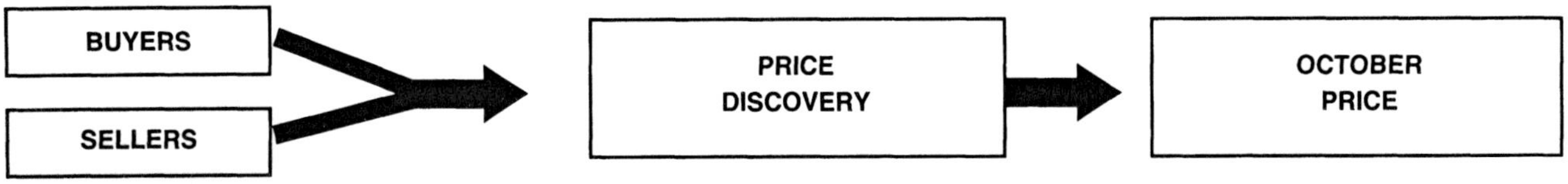

Mix buyers, sellers, and information and get price discovery

It is very important that we understand that the system **needs** to discover a price for October based on information that is available in early April. But it is also becoming apparent that this price expectation is likely to change as we go through the planting season, find out how many acres are actually planted to soybeans, move into June and July during the critical growing period, see what type of weather we have and what type of yields per acre are likely to be generated, etc. Thus, price expectations will change over time as the information based on supply, in particular, changes. We would logically expect those types of price adjustments to occur, and the futures market will in fact discover different prices as the weather outlook changes.

Weather can change price expectations

PRICE DISCOVERY IS DYNAMIC AND UNCERTAIN

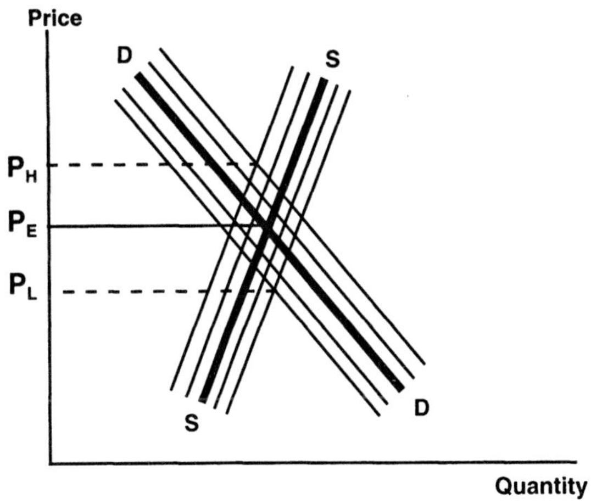

Graph 2 - Demonstration of various prices for soybeans that can occur

Even on April 10, therefore, it is not a precise scientific process that is discovering a price for the future time period. Graph 2 provides a useful frame of reference. It shows a band of demand curves and a band of supply of curves with bold demand and supply curves intersecting at the equilibrium price that was discussed in Chapter 1. There is a price that will balance what the demand for the product will be in early October and what the supply of soybeans will be in early October, but no one knows on April 10 precisely what that price will be. Thus, buyers and sellers bring their *expectations* to the market, and as Graph 2 suggests, transactions may occur at prices as low as P_L and as high as P_H.

At this point, the beginner may be getting a little frustrated. One can almost hear the reaction that goes something like, "See, I told you this marketing stuff was complicated. Apparently, we cannot even know for sure what the price is going to be!" But the reality is that the pricing process is imprecise, the knowledge base is not exact, and that interpretation of existing information on demand and supply will differ across different market participants or different traders. *That having been said, it does not mean that there are no dimensions of price movement over time that are reliable and useful.* Understanding this last point can turn the farmer who will persevere into a tough competitor and a profitable operator over time.

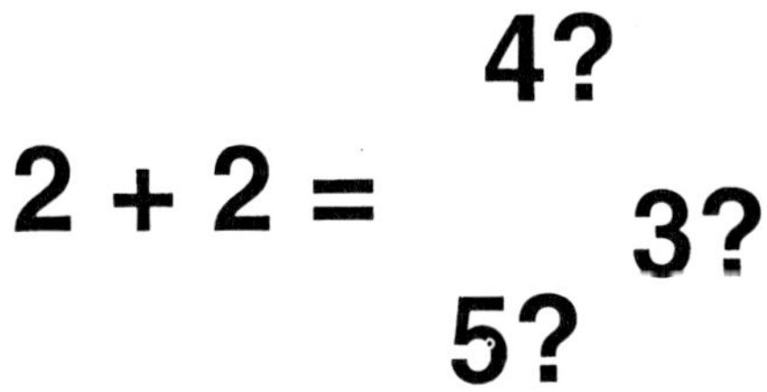

Marketing is uncertain, but there is still logic

SEASONAL PRICE PATTERNS

Almost every agricultural commodity has a seasonal or within-the-year pattern in prices. It is very important that the farmer know that this pattern exists and know how prices in one month tend to compare to prices in another month. The seasonal price patterns can be predicted. The reason the pricing patterns are predictable and reliable and tend to recur over time is because they are tied to relatively consistent seasonal patterns in production. It will be important to look at these pricing patterns and develop some management tools that are related to the patterns for both storable and non-storable commodities.

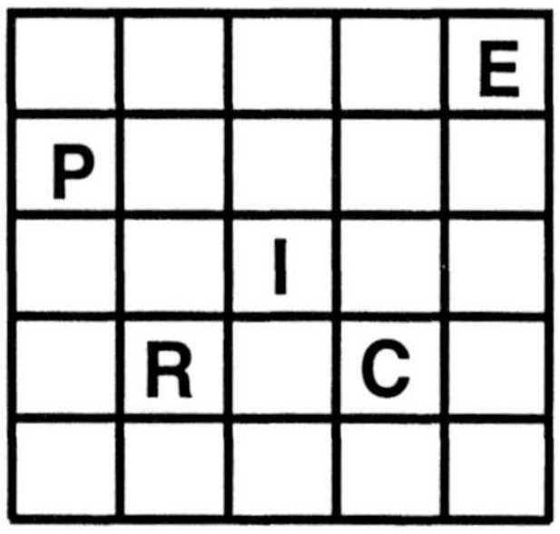

There is a price pattern here someplace

STORABLE COMMODITIES

The storable commodity tends to show a price pattern like the one sketched in Graph 3. What is assumed is that price will be lowest in and around the harvest period when there is the biggest quantity of product available to the market. Then, price will tend to rise reflecting storage costs as we move through the year. The price increases do not typically continue until the next harvest, however. If it is wheat, for example,

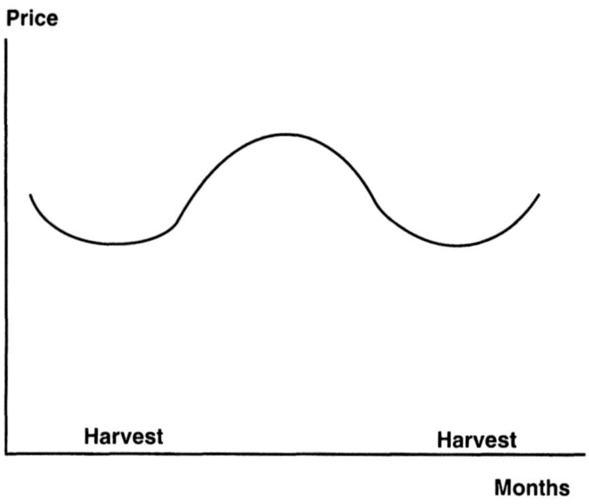

Graph 3 - A typical seasonal price pattern for a storable agricultural commodity

and the bulk of the winter wheat is harvested in the U.S. in June, prices will not be low in June and July of year 1 and then increase all the way through May in year 2. In March, April, and May, as the next harvest approaches, many potential buyers or users of wheat can start to delay their purchases and wait on the next harvest. They may have sufficient stocks to cover their immediate processing needs until the next harvest arrives when somewhat lower prices might then be available.

Production patterns are the key

In a normal pattern, therefore, whether it be wheat, corn, soybeans, barley, cotton, or anything that is storable, prices tend to be lowest during the harvest period. They show a tendency to increase following harvest reflecting storage costs, and then start to decline again as the next harvest period approaches.

There can and will be exceptions to this pattern. The exceptions usually come when something abnormal occurs. Graph 4 plots monthly average corn prices in central Illinois for the crop years 1991-92, 1992-93, and 1993-94. The "crop year" for corn starts September 1 and runs through the following August 31. Note that the monthly prices for late 1993, which would be early in the 1993-94 crop year, do not show the lows at harvest like those seen in and around October of 1992. This is because the widely publicized floods in the Midwest in the summer months of 1993 damaged the corn crop and pushed the crop from an initial estimate of over 8.0 billion bushels down to 6.344 billion bushels. This sharp reduction in production moved the supply curve for corn up and to the left along a negatively sloping demand surface.

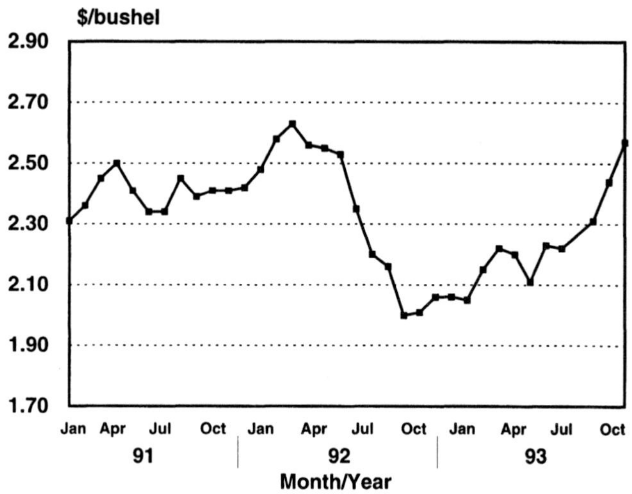

Graph 4 - Monthly corn prices, central Illinois, 1991-1993

As we noted in Chapter 1, this decrease in supply means prices of corn will go up. Prices in and around harvest in the fall months of 1993 were above prices that were being registered early in the calendar year. Indeed, after the November report by the USDA confirmed the sharp reduction in production, prices for November, December, and early 1994 were as much as $.80-$1.00 per bushel above year-earlier levels.

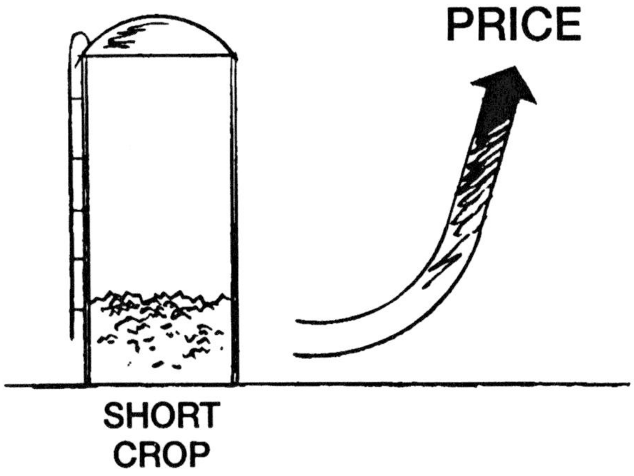

A short crop means tall prices

It is also important to recognize that the anticipated increase in prices following harvest may or may not be sufficient to cover storage costs. The decision maker has to step back and use market indicators to determine whether or not corn or any other storable commodity ought to be sold at harvest or stored. Some of those market indicators and criteria will be developed in later chapters dealing more specifically with price risk management strategies, but it is important to recognize that the magnitude of the within-year price increase is somewhat unpredictable. Prices will almost always go up in the months following harvest, but they may not go up enough to cover your storage costs.

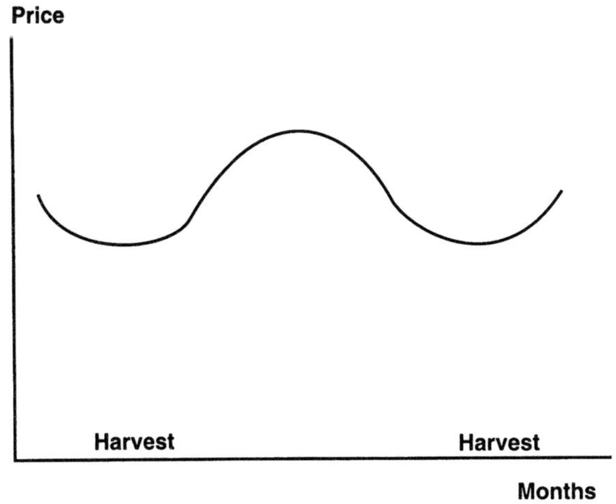

Graph 5 - Price increases may not be enough to cover costs

No profits guaranteed

Keep in mind that what the marketplace will do is give some incentive to hold storable commodities so that those commodities will be available for processors all during the year. It gives that incentive in the form of possible price increases after harvest. That is the function of the marketplace. When it discovers prices that are fairly low at harvest and shows the potential for somewhat higher prices six months later, it is encouraging some producers to not sell at harvest and store the commodity so it will be available six months later. But given the competitive status of agriculture, one would not expect every person who stores corn, soybeans, or cotton to profit from their storage program. *All the marketplace should have to do is make it profitable for the more efficient producers.* That usually is enough to encourage storage and spread the product through the year and make it available on a year-round basis.

The marketplace rewards the most efficient storage program

Bin Cost is not variable

There is a message here. Historically, many farmers have felt that if they had storage bins on the farm, they absolutely had to use them to store their corn, soybeans, or wheat. That can be a serious mistake. If the bins are already built, *there are no variable or controllable costs associated with their use. They cost the same whether they are used or whether they are empty.* It is very important to recognize that in some years, the astute producer who has become an astute marketer will let the bins sit empty and will sell the commodity at harvest time! We will come back to this storage program in considerable detail in later chapters. *It is very important to learn to do this right.*

It can be a mistake to insist on using storage bins

What gives the storable commodity the seasonal price pattern, then, is the seasonal pattern in production. Wheat is produced in the U.S. both across the winter and summer growing periods, but most of the wheat is sowed in the early fall months and harvested in the following June or July. Thus, the harvest period for wheat is June, and prices tend to be lowest then. For corn, the typical harvest month is October. For soybeans, it can be either October or November. For cotton, it is November.

The primary point, before leaving this topic, is that seasonal patterns in production are what prompt the seasonal patterns in prices for storable commodities. The patterns in prices are therefore reliable, predictable, and will tend to exist over time. Since they do, the producer can take advantage of them.

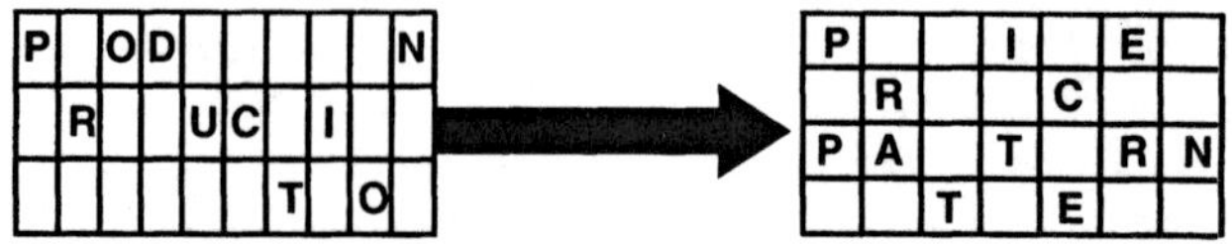

For storable crops, the production patterns causes the seasonal price pattern

NON-STORABLE COMMODITIES

Fresh fruits are just one example of a commodity that may not be storable or is only semi-storable. The livestock commodities are an example of commodities that are not storable. Meats can be frozen and kept in storage, but there is a small amount of storage in the U.S. for most meats. It is small compared to total supplies, and is usually restricted to a few select commodities such as frozen pork bellies that are sliced to make bacon. Usually, hams are sold fresh or processed, not frozen. There is an attempt to sell most poultry, especially chicken, fresh because most consumers would prefer it fresh and not frozen. Even the tendency to freeze turkeys, which was historically a characteristic of the market, is starting to change. Many turkeys come to market around the Thanksgiving and Christmas holidays chilled and ice-packed but never frozen. The seasonal price pattern for such non-storable commodities, we will find, is also related to the seasonal pattern in production.

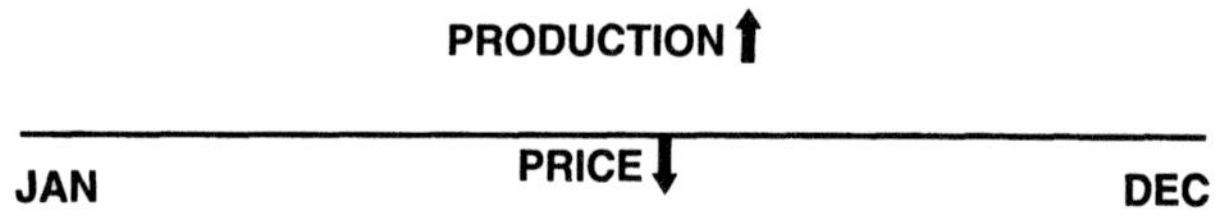

When the production season hits, prices drop

Graph 6 plots cash prices for 1991-93 for slaughter hogs. These prices tend to show strong seasonal patterns. There is usually a period of price weakness in March and April, and then prices get pushed lower in October and November. These periods of price weakness occur because many hog producers try to avoid either the extremely hot months or the extremely cold months for the birth of their pigs, called "farrowings." Most producers would prefer to farrow in the fall or spring months. It turns out that in a well-managed production program, hogs can reach slaughter weights in six months. Thus, when there is surge in farrowings in the spring months, there will be a predictable surge in slaughter of hogs six months later, essentially October and November. The same seasonal increase in slaughter occurs in the spring months, six months after the surge in fall farrowings.

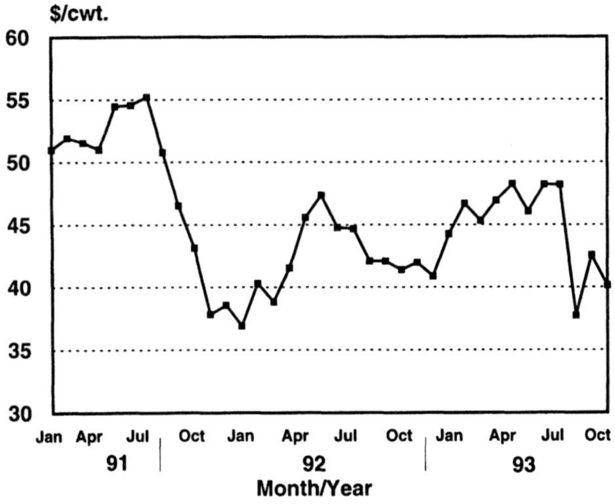

Graph 6 - Monthly hog prices for 1991, 1992, and 1993 to show the persistent seasonal pattern in prices

The hog cycle will change things

The seasonal pattern for 1991 is "tilted," with bigger price declines late in the year than for 1992 or 1993. The 1991 year was also being influenced by the cycle in hog prices. Earlier favorable prices had prompted producers to expand, and that increased the supply of hogs significantly by late 1991. Remember, one of the common dimensions across commodities discussed in Chapter 1 was the need to be aware of what other producers are doing. The 1991 "break" in hogs prices is a classic example of the price pain a producer can feel because other producers expanded too much and shifted the market-wide supply curve to the right. We will come back to the "price cycle" later in this chapter.

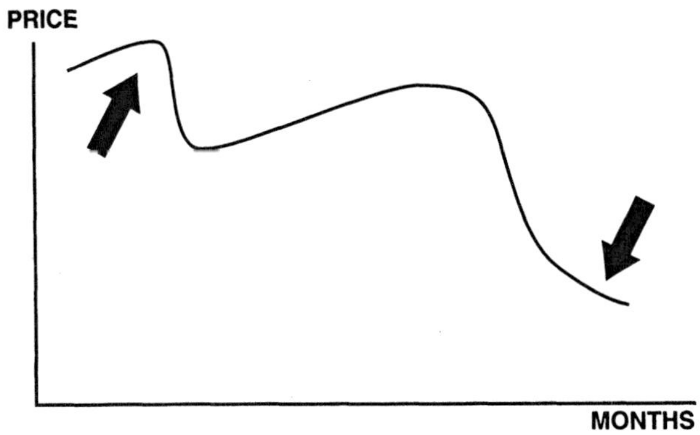

Graph 7 - The price cycle can "tilt" the seasonal pattern

It is still the production patterns that move price

Graph 8 shows monthly slaughter of hogs across a recent three-year time period. Note that there is substantial variation in slaughter levels as you move through the year. Part of this, of course, is due to different numbers of work days during the month. This is one of the reasons that February tends to be a relatively low slaughter month. The seasonal pattern is still clearly there, however, and is consistent with the notion that we have already established. *In the short run, it is changes in supply that drive changes in price.* Remember that in months in which slaughter tends to be high, prices tend to be low.

When supply increases, price decreases

What is happening, of course, is that the monthly supply curve is moving along the negatively sloping demand curve that we talked about in Chapter 1. Price, in order to clear the market, does the adjusting. When slaughter levels increase in October and November, the supply of pork increases. On a month-to-month basis, demand does not change much so the demand curve is largely stationary. The monthly supply of pork increases, the supply curve shifts to the right, and consumers take the increased quantities but only at a lower price. This is all very logical, very basic, and very important.

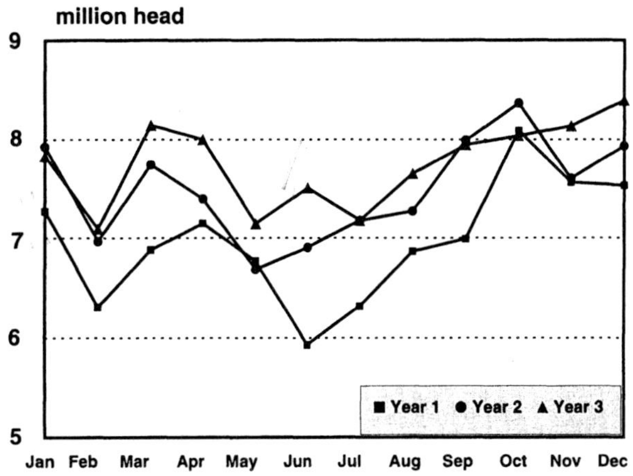

Graph 8 - Monthly hog slaughter levels for three representative slaughter years

Seasonal patterns in cattle prices will be somewhat different than those in hogs. The production pattern is different. Slaughter levels tend to be higher in the fall months. Cattle are not only coming out of the feedlots where they have been fed on concentrated grain rations, but cattle are also coming to market near the first frost in the fall when the grass stops growing. There are many types of cattle programs that are based on grazing pastures during the spring, summer, and early fall months, and when frost occurs, that often is the first indication that cattle must go to market. A normal seasonal pattern in cattle, then, shows the best prices in the late spring and early summer months when the grazing period is best and there are fewer cattle coming off pasture to market. Then, prices tend to be lower in the fall when that increased slaughter occurs. The seasonal patterns in cattle are seldom as pronounced as they are in hogs because the differentials in monthly slaughter levels are not as great.

What decision makers need is a tool to let them take advantage of their knowledge of reliable and predictable price patterns within the year. Fortunately, such a tool is available and it is very simple to calculate and very simple to use. It is called a **seasonal index.**

An index helps us measure

A seasonal index is made up of 12 percentages or index levels, one for each month, and the 12 average 100. *The index value for any particular month will show what the price will normally be in that particular month compared to the yearly average price.* Let's use the previously discussed hog situation to illustrate. In discussing the monthly prices in hogs, mention was made of the tendency for prices to be relatively low in October and November. If the seasonal index for October is 94, this means that the October price will tend to be only 94 percent of, or 6 percent below, the yearly average price. Conversely, the seasonal index for July may be 107. This suggests that July prices are normally 7 percent above the yearly average price. As suggested, there will be an index, a percentage, for each of the 12 months.

JANUARY	FEBRUARY — — — —	DECEMBER
98	104	98

A seasonal index is a measure of price for each month

DOING A SEASONAL INDEX

There are a number of relatively sophisticated ways of calculating a seasonal index. Some analysts use a moving average process, designed to eliminate the influences of longer-run price movements such as a price trend and to isolate the true seasonal pattern. A somewhat more simplistic, but nonetheless very effective approach, is to take the most recent 5-10 years and average the prices for each month across the time period. For example, if you have 10 years of data, all 10 January's should be added and then averaged by dividing by 10. Do the same thing with February through December. Once the 12 monthly average prices across the 10-year time period have been calculated, then add those 12 monthly numbers and divide by 12 to get the average price across months. Set that price to equal 100 in a percentage or index context, and then calculate the index values for each month. To make sure that this is clear a sample is shown in Table 1.

Assume average hog prices across 10 years are as follows for each month:

January	49.00	July	54.00
February	52.00	August	53.00
March	47.00	September	51.00
April	47.00	October	47.00
May	50.00	November	48.00
June	53.00	December	49.00

The average price across months in this simple demonstration would be $50.00 per hundredweight. Thus, the index value for each month would be:

January	$49.00 ÷ 50.00 = .98 or 98%
February	$52.00 ÷ 50.00 = 1.04 or 104%
December	$49.00 ÷ 50.00 = .98 or 98%

Table 1-calculating a seasonal index

The lowest index number for any of the months shown would be 94 percent ($47.00 ÷ 50.00) for March, April, and October, each with the historical average price of $47.00. The highest index would be for July, 108 percent ($54.00 ÷ 50.00). If you take all of those percentages, after calculating one for each month, and add them and divide by 12, the average would be 100 percent. *Thus, we have a seasonal index that shows how each month would compare in a normal year to the average price for the year.*

Table 2 shows seasonal indices for a number of widely produced agricultural commodities. They are reflective of the seasonal patterns that one would expect given the earlier discussion about storable and non-storable commodities. Milk was not discussed specifically, but one would expect to see monthly milk supplies and production be the greatest in the late

Month	Corn	Milk	Soybeans	Wheat	Hogs	FedCattle	FeederCattle	Calves
				COMMODITY				
January	98.92	102.82	97.77	101.80	97.94	94.85	101.00	94.94
February	99.37	101.78	97.36	102.96	99.30	97.64	101.80	101.54
March	100.67	100.68	100.11	102.77	95.33	100.94	100.30	102.48
April	102.78	98.28	102.61	101.98	93.56	103.68	99.00	105.05
May	104.69	97.21	105.12	100.83	100.42	103.13	97.60	106.17
June	105.81	96.78	105.10	97.98	105.09	101.40	98.40	101.51
July	105.98	96.86	103.55	97.49	107.25	99.49	100.20	98.69
August	100.18	98.30	101.08	97.53	106.40	99.29	101.20	99.25
September	96.24	99.88	100.01	98.15	103.64	101.46	100.00	99.47
October	94.40	101.73	95.92	99.25	99.55	101.52	99.00	97.84
November	94.52	102.58	95.83	99.47	95.11	98.64	100.50	96.24
December	96.44	102.31	95.56	99.79	96.41	97.96	101.80	96.80

Table 2-Seasonal index for typical agricultural commodities

spring and early summer months when pastures are plush. This is, in fact, reflected in the seasonal index for milk. Even though it is a government program commodity, there is still some seasonal variation in prices, and milk prices do tend to be lowest when the monthly supply is the greatest in the late spring and early summer months. The patterns for most of the other commodities fit intuition and what one would expect given the issues that relate to the production patterns we have already discussed.

USE OF THE SEASONAL INDEX

Once the seasonal indices are generated, they can then be very useful to the decision maker interested in analyzing various marketing decisions. *The indices can be used to predict a price for some month later in the year.*

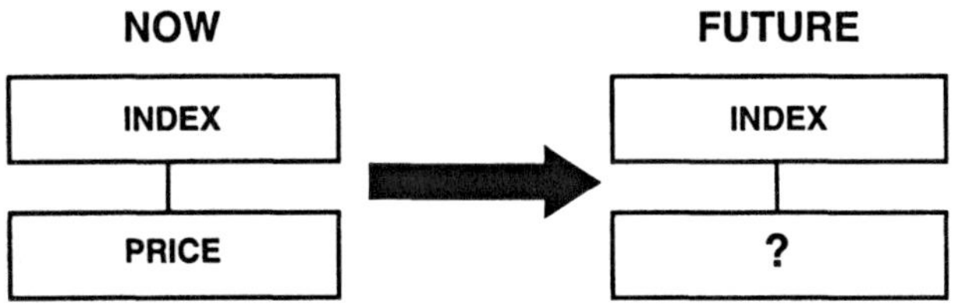

The index value can provide a price prediction

Let's assume, to illustrate, that we are dealing with the same set of simplistic averages generated in Table 1 and it is April. In April, you know the seasonal index for hog prices is 94 or 94 percent. If you know that current prices in your market area are at $44.25 per hundredweight (you called the local meat packer), then you have a basis for predicting what hog prices will be at some later time, say July. *This ability can be very important in evaluating a contract price that is being offered to you by the local packer or deciding whether now is an appropriate time to think of using some of the various techniques that we will get into later to forward price the hogs.* Currently, what we want to do is show how this index process can be used to predict prices. An example is laid out in Table 3.

April index: 94
April price: $44.25 per hundredweight
July index: 108
July price: ?

Let X be the unknown July price. Set up

$$\frac{\text{April Index}}{\text{April Price}} = \frac{\text{July Index}}{X}$$

and solve for X. Procedurally, we have

$$\frac{94}{\$44.25} = \frac{108}{X}$$

$$94X = 108(\$44.25)$$
$$94X = \$4779$$
$$X = \$50.84$$

Note the math is quite simple. Once the ratios are set up, we can solve for X by cross multiplying to create

$$94X = (108)\ \$44.25 \text{ or } \$4{,}779$$
$$\text{Then, } X = \$4779 \div 94$$
$$= \$50.84$$

Table 3-Use of a seasonal index to predict a later price

You must be able to evaluate price

To predict price

Note that once the price of $50.84 is generated as an estimate for July price, the decision maker is then in a good position to evaluate other pricing opportunities that might be available. Assume, for example, that the local packer who is interested in trying to get hogs committed to make sure he has adequate slaughter levels in July is offering a price of $48 per hundredweight for early July delivery. As a farmer, it is important to be in a position to evaluate that $48 offer compared to where you think the prices are likely to be. The $48 looks good, for example, compared to $44.25 that is currently being paid in April. But using the seasonal index suggests that you are quite likely to see prices substantially better than the $48 in July. That knowledge can be used to turn down the offer from the packer or to try to negotiate a still better contract price once the buyer learns that you do, in fact, understand the seasonal patterns in hogs prices.

Being able to predict hog prices will help

When we start talking later about other types of forward pricing possibilities, such as using futures or options on futures, this ability to predict within-year price movements will again be important. For example, in April when this estimate of $50.84 is generated, we look in the newspaper or otherwise access what is going on in the futures market and find that the futures complex in Chicago is discovering a price for July of $53 per hundredweight. Given your prediction of $50.84, this starts to look attractive, and you may decide to follow through on getting those hogs forward priced at the $53 price. Actually, the astute reader might step back from this and say, "If the futures market is discovering a $53 price, why is the local packer offering only $48?" That is an interesting question, and in actuality those two prices (futures vs. packer offer) are not likely to be as far apart as the prices used here. What we will find is that the price that the packer is willing to offer is probably tied to what is going on in the futures pits in Chicago. But we will come back and develop that in more detail later. *The important point here is to recognize that having the seasonal index at your disposal puts you in a position to have a very useful piece of information in terms of understanding price patterns within the year.*

A seasonal index helps in making storage decisions

To help in the storage decision

The seasonal index is equally important and equally useful in helping make other decisions that many farmers often find difficult. An example is the storage decision. Let's assume that we have a farmer who is harvesting wheat, it is June, and prices are $3.00 per bushel. We know that the seasonal index for wheat in June is 94, which means that at the peak harvest month, prices are normally 6 percent below the yearly average price. You are interested in storing the product until December, but you want to sell it in this calendar year for tax reasons. So you would like to know what type of price increase you are likely to get between June and December. You look at the December seasonal index and find that it is a 105. This creates an opportunity for you to predict a price for December and to estimate how you will fare if you sell the wheat in June as compared to incurring storage costs and store it until December. This example is laid out in Table 4.

June:	June index	=	94
	Cash price	=	$3.00
	December index	=	105
	December price	=	?

Set up $\frac{\text{June Index}}{\text{June Price}} = \frac{\text{December Index}}{X}$

where X = the unknown December price. The math becomes

$$\frac{94}{\$3.00} = \frac{105}{X}$$

$$X = \$3.35$$

You just cross multiply and solve for X as demonstrated in Table 2.3.

Given an estimate of $3.35 for December, let's evaluate to see whether it is better to sell in June or hold until December. Do the following:

June price	$3.00
plus Cost of storing until December	.18
equals Break-even in December	$3.18

Table 4-Use of seasonal index to evaluate the store-sell decision in wheat

This suggests you will need $3.18 in December to be equal to selling at $3.00 in June since it will cost an estimated $.03 per bushel per month or a total of $.18 to store. The tentative decision: store the wheat and sell in December.

The decision comes out to store the product and sell it in December at the expected price of $3.35. *There is a caution that is useful here.* The $3.00 that you can get in June is a sure thing. You can haul the wheat to the local grain elevator later in the day, and the elevator manager will pay you $3.00 per bushel because that is the price that he has bid for your particular quality of wheat on that day. *The $3.35 price and the added profit from storage ($3.35 - 3.18) or $.17 per bushel compared to selling in June in December is not a sure thing.* Any number of things can happen that can cause the price in December to actually be below $3.18, the price needed to make storage equally profitable with selling June. Assume, for example, that in the late summer months, the Wheat Marketing Board in Canada decides to sell a lot of stored wheat on the world market. Prices are driven lower. What we see, then, is that there is a risk associated with holding the wheat and hoping for the better price in December.

SUBSTITUTE | **SURE PRICE** | **FOR** | **CHANCE OF HIGHER PRICE** | **?**

This one needs some careful thought

Actually, you can "have your cake and eat it too." What we will find in later chapters is that when you work out this analysis and find you can get a $3.35 price in December, then it is better to store the wheat. *What you need to do is go ahead and forward price and establish that $3.35 price.* The reasons to be concerned about what the price will actually be in December are then eliminated. You can hold the wheat in storage and anticipate added profits compared to selling at harvest with a considerable amount of confidence. All this is possible, note, because of the use of this very basic and very simple seasonal index. It helps you understand and appreciate how and why prices move within the year. *It is a powerful marketing tool.*

Tying down that future price can help

PRICE CYCLES IN AGRICULTURE

Most agricultural commodities, as we have seen, show significant seasonal price patterns. The other big supply-side mover in prices, one that extends across years, is the price cycle. Understanding of these two typical and reliable patterns in prices of agricultural commodities, especially the livestock commodities, is very important to the decision maker at the farm level.

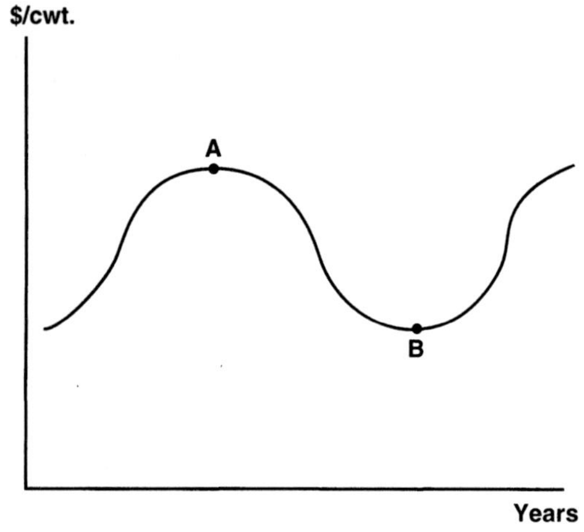

Graph 9 - A typical price cycle in slaughter hogs

The typical price cycle plot looks like that shown in Graph 9. On the vertical axis are prices, dollars per hundredweight, for slaughter hogs. On the horizontal axis is time, typically in years, when we are looking at a price cycle. Graph 9 shows the tendency for prices to sweep up, then down, and then sweep up again. Obviously, producers would like to know when they are at a point such as A on the graph and prices are headed toward substantially lower levels. Conversely, and for somewhat different reasons, producers would like to know when they are at point B and prices are expected to climb in a sustained fashion for several years. *It is obvious that understanding the price cycle, why it occurs, and what affects its length is going to be important.* Not only will this help make production-level decisions better, but it is a necessary base for understanding management of price risk to be discussed in later chapters.

The cycle pushes prices up and down over time

Obviously, there are reasons why prices are high in the area marked by A in Graph 9. There are reasons why prices are low in the area marked by B. The place to start in discussing the cycle, then, is with why we have a cycle:

- **What is it that causes prices to move higher for a sustained period and then move lower for a sustained period?**
- **Is the underlying behavior behind these patterns predictable and reliable or is it more of a random process?**

SUPPLY IS THE MAJOR PRICE MOVER

It is important to recognize that the high prices near A occur because the supply for this particular agricultural commodity has been reduced. We have the same phenomenon that was discussed earlier in talking about seasonal price patterns. For some reason, the supply has been shifted to the left and up a negatively sloping demand curve. This means that the price goes from a level such as PL in Graph 10, up to a level such as PH. In other words, prices go from relatively low levels to relatively high levels because supply has decreased--the supply curve has shifted to the left.

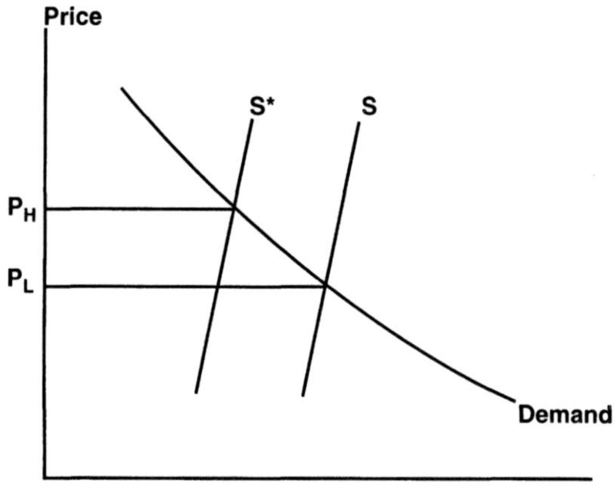

Graph 10 - The supply shifts that prompt price changes in a price cycle

Walking through a cycle

To understand what occurs in price cycles, it is useful to think through what goes on periodically and frequently in the swine market. If you break into the price cycle at the point at which prices are starting to increase because daily or weekly slaughter levels are starting to decrease, it is then useful to think about what the individual producer is seeing and thinking. Let's assume that we have a fairly efficient producer who can cover all costs in hogs at $42 per hundredweight.

We better be careful: everyone will try to expand with these prices

Prices start to go up, and move above $50 per hundredweight. Prices at this level are extremely attractive. The typical producer is going to start thinking about expanding so as to have more hogs to offer at what looks like very favorable prices. To expand the breeding herd, gilts (unbred females) have to be diverted from the feeding floor and into the breeding program. This tends to reduce the number of hogs available for slaughter, which shifts the supply curve for slaughter hogs to the left even further, and prices are kicked still higher. Now, as prices approach, say, $55 per hundredweight, the farmer down the road starts to get the urge to expand, and the process is repeated on his farm. Gilts are pulled off the slaughter channels and into the breeding herd, prices are pushed still higher, and still more people feel the strong motivation to expand and get some of those higher prices. *This process is like the snowball rolling downhill--it builds and feeds on itself.*

Don't get caught in front of that snowball

Eventually, the breeding herd grows to the point that the available supply of slaughter hogs starts to increase. From the time that a gilt is bred until the pigs from that first litter reach market weight is roughly nine months. There is a 114-day gestation period from the time of breeding until the pigs are born, and then a good producer can have those pigs to slaughter weights, in the 230-250 lbs. range, in six months or less. What this means is that after several months, there is an expanded level of slaughter coming from the decisions of the early producers, the producers who decided to expand at the first signs of higher prices. As a few more weeks and months pass, even more of the hogs that were coming from the expanded breeding herd are now ready for slaughter. Slaughter levels on a daily and weekly basis start to increase, often dramatically.

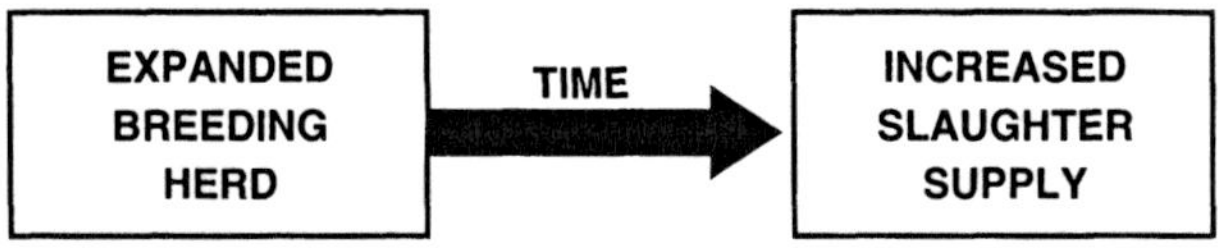

It's just a matter of time

Expansion and liquidation phases

At this time, the point A in Graph 6 has now been passed. Prices have peaked and are starting to go down under the weight of increased slaughter. What is happening in Graph 10 is that the supply function is being shifted from the one marked S* to the one marked S, and, as would be expected, price will start to fall from relatively high levels, the PH level, down toward the PL level. As this process continues and as prices start to decline, some producers, especially those who lack financial strength, will start to panic. They may even sell their breeding animals. This starts the **liquidation phase** of the hog cycle as compared to the earlier **expansion phase**. Sows that have been in the breeding herd start to come to market. Occasionally even bred gilts will come to market, and the supply of slaughter hogs increases. Prices are pushed sharply lower by this process as each successive producer has to make a decision on whether or not to stay in the hog business. It is not at all unusual to see prices during this liquidation phase fall substantially below even the most efficient producer's cost of production.

The liquidation phase of the hog cycle

Eventually, however, even this process runs its course. The breeding herds get reduced to the point that the base potential for producing slaughter hogs has been significantly decreased. The supply of hogs coming to market on a daily or weekly basis starts to decline and prices start to increase. Some producer somewhere starts to feel that urge to expand as prices climb back up toward $50 per hundredweight or better--and the process starts all over again. Historically, about every three to four years the swine industry in the U.S. has gone through this process of over-expansion, forced liquidation, then over-expansion again. Prices move up and down in a rhythmic, repetitive, and fairly predictable cyclical pattern.

Shown in Graph 12 are quarterly average hog prices since the mid-1970's. The prices are volatile, but the rhythmic and repetitive cyclical pattern occurring about every three to four years is certainly there. Look, for example, for the periodic highs in prices or the periodic dips in prices, and note that they are often three to four calendar years apart. The most recent cyclical lows in prices come in quarter 3 of 1988 and quarter 1 in 1992, a 3.5-year cycle. Prices were better in 1993 and early 1994 but plunged below $30 in late 1994. The cyclical liquidation started in late 1994 shows every sign of being a severe one.

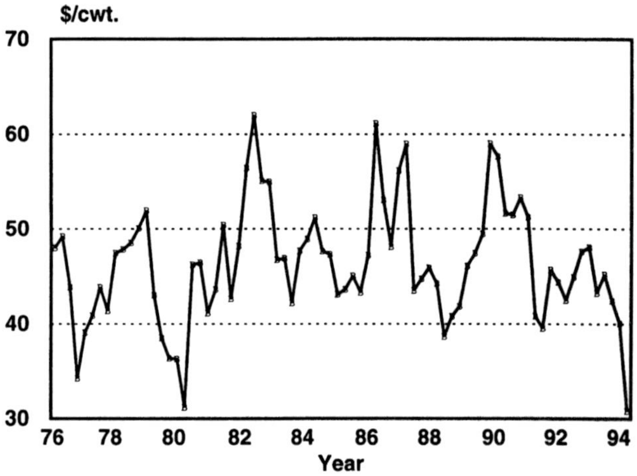

Graph 12 - Hog price cycle: quarterly average slaughter hog prices, 1976 to 1994

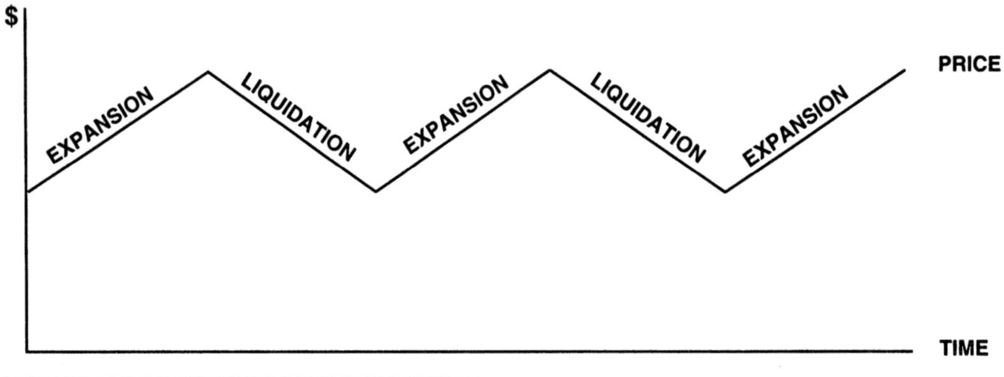

Graph 11 - The hog cycle is completed every three to four years

Examination of Graph 12 helps the producer understand how the industry is changing. It appears that the magnitude of the price swings from peaks to troughs is not always as great as it was in earlier years. This is probably true, and it has occurred for a good reason. In the 1970's and early 1980's, there were a lot of hog producers who produced hogs as one enterprise on a farm. Often, there was relatively little investment associated with that enterprise. The sows may have farrowed the pigs in open sheds with little climate control. Slaughter hogs often ran in a fenced lot and came up to a feeding trough with very little machinery, buildings, and equipment associated with what was often a small swine enterprise. When prices dipped below cost of production, these types of producers could sell the few sows or gilts that they were holding and get out of the business until prices started to look better again. This in-and-out process occurred because those same producers might come back into production when hog prices start to climb back to levels above their cost of production.

With all this money invested, you have to operate

The industry is changing

As the industry has changed and more of the hog production is in the hands of relatively large year-round producers, the cyclical patterns seem to have dampened. The modern producer may have 1,000 sows or more, and all of the hogs are under fairly carefully controlled conditions and confined to modern buildings with high-tech capability for handling manure, waste, and other by-products of a swine program. This type of producer has a substantial financial investment. It is very costly to just stop producing hogs and absorb the annual mortgage payments on that investment. The buildings, the farrowing crates, and the equipment have to be paid for. Thus, modern producers seldom go in and out of business.

Managing the cycle

What they may do, at most, when they spot an approaching price decline as suggested around point A in Graph 9, is sell some of the older sows while prices are still high or sell some of the sows that are weaning a smaller number of pigs. Thus, they "trim up" their operation to keep the best performing and most efficient sows and try to operate at a level that best fits the buildings and other accommodations. Later, when prices are booming, they may crowd those buildings and carry some sows that are only weaning 9 or 10 pigs instead of the 14-15 that might be coming from the better sows. They may even hire part-time labor to help handle an expanded operation during the period of booming prices. *But they make no permanent long-term changes in terms of their production capabilities.* They just "manage" the cycle by changing the number of hogs that are produced depending on whether prices are expected to be up or down.

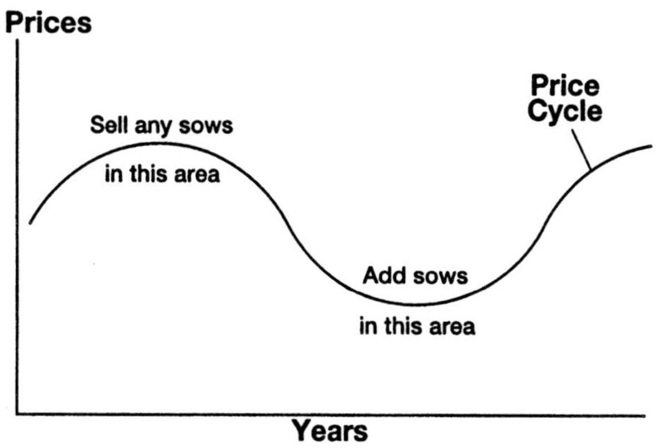

Graph 13 - Cull sows should be removed when sow's prices are high

Later in chapters dealing with price risk management, we will return to this point. Obviously, it is very important at a point such as point A in Graph 9 to think about getting hogs forward priced into the period where prices will be low, around point B. There is no foolproof way for a producer to escape all of the price pain resulting from the hog cycle, *but much of it can be avoided by an effective and aggressive program of forward pricing.* We will deal with this explicitly and specifically in later chapters on price risk management.

Before leaving the discussion of price cycle in hogs, it is useful to emphasize the point about supply changes bringing most of the changes in price. Graph 14 repeats the quarterly hog prices shown in Graph 12 but also adds quarterly pork production as a measure of supply. This graph looks complicated, but it is worth some study. It has two scales, price on the left vertical scale and production on the right vertical scale. It illustrates the importance of supply shifts. *Note that in general when production is up, price is down, and when production is down, price is up.* This is especially apparent when production changes in a significant manner. This is just another way of presenting the fact that the supply curve is moving along a demand curve as was suggested in Graph 10. Supply changes as producers respond to prices, both high prices and low prices, *and it is those supply changes translated into production changes that cause prices to move in the opposite direction.*

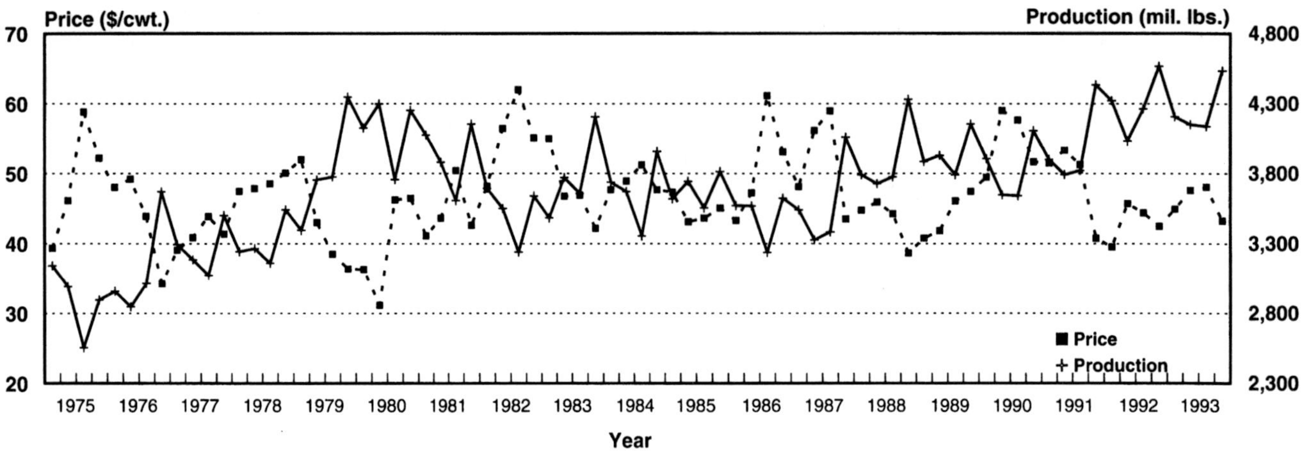

Graph 14 - Hog prices and pork production by quarter, 1975 to 1993

There is a behavioral dimension

Some aspiring swine producers might be wondering at this point why the cycle appears to be three to four years in length. After all, we have already indicated that from the time a gilt is bred until an expanded number of slaughter hogs can be made available to the market is only nine months or less. We have to keep in mind that it is not just the biological dimension of the cycle that determines its length. Every producer has to **decide** to expand or **decide** to liquidate before actually taking those actions. What happens, then, as prices start to move up, is that the expansion comes with some time lag. It tends to build and build until eventually the supply that is ready for the market is too big and starts to push prices down, and then the final acts of liquidation come, again with some time lag. We might argue, therefore, that there is both a **biological** and a **behavioral** dimension to the hog cycle. If you add the biological part to the behavioral lags, the time it takes for producers to decide to make up their minds, then you will find that it is typically three to four years as we move from price peak to price peak.

TIME TO DECIDE + **BIOLOGICAL TIME LAG** = **LENGTH OF CYCLE**

Making decisions affects the hog cycle

Careful examination of the production and price plots in the mid-1980's indicates that sometimes cycles tend to be longer than the expected three to four years. There were times in 1986-87 when hog producers were not quick to expand in the face of very profitable prices, prices that surged above $60 in 1986. They were coming off of some difficult and unprofitable times earlier in the decade, and when their enterprises started to become more profitable, they paid off debt instead of expanding their hog programs. They were not going to the bank to borrow money. Rather they were trying to reduce their debt load as hogs paid them good profits during 1986 and into 1987 (review Graph 12). Analysts and producers who were watching the market and expecting a cyclical expansion and a subsequent price break had to wait an extra year before it occurred. We need to keep in mind that cyclical analysis of markets is not an exact science. The cycle involves people and their decisions to expand, to cut back, to act, and people are never totally predictable. This is especially true in volatile situations like the one faced by the typical hog producer.

This time, hog producers reduced debt with their good profits

A longer cycle in cattle

Cycles occur in many agricultural commodities, not just in hogs. This is especially true in the livestock sector. Graph 15, for example, shows a plot across a number of years of the January 1 inventory in the cattle sector. Going back toward World War II, numbers show a tendency to surge, decline, surge, and decline again in about a 10-year cycle. There has been considerable talk over time about a tendency for cattle inventory numbers to peak every 10 years. Note the peaks in 1955, 1965, and 1975.

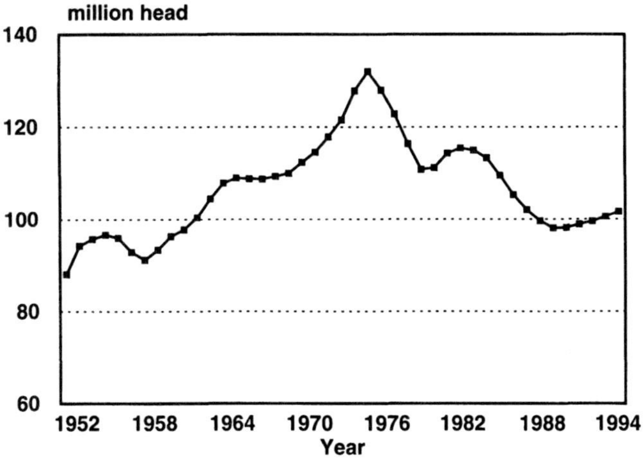

Graph 15 - January 1 cattle inventory numbers, 1950-1995

What occurs is the same sort of phenomenon as the one discussed in hogs. The cow herd has been reduced by some past actions, let's assume, and the prices of the weaned calves the producers have to sell start to increase. When those calf prices climb well above the producers' estimated cost to produce that calf, the urge to expand starts to develop. In the beef cattle business, you expand by holding back a heifer (unbred female) and getting her ready to go in the breeding cow herd instead of allowing her to move into the slaughter channels, or into the feedlot complex and then into slaughter channels. The gestation period for bred beef cows is 280-285 days, so it takes much longer to expand the available supply of slaughter animals in beef than it does in hogs. Instead of having litters of 10-15 pigs, the brood cow has one calf per year. That biological process cannot be speeded up a great deal, and typically the calf is a year or more old before it is sold for the first time.

When you add this up, it takes anywhere from five to seven years for the expansion phase of the cattle numbers to run its course. Then, comes the liquidation phase, a phase in which producers are so stressed because of low prices that they sell cows out of the breeding herd or sell bred females. This liquidation phase is typically somewhat shorter than the expansion phase, more in the magnitude of three to five years, and the two when combined have created the much-discussed and traditional 10-year cycle in cattle.

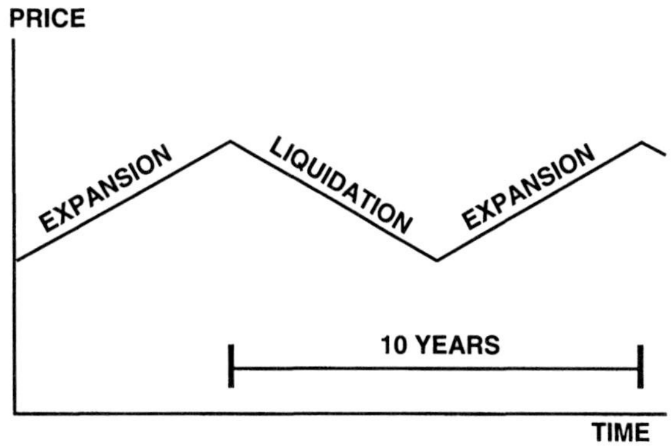

Graph 16 - The cattle cycle can take 10 years to complete

The last peak in numbers came in 1975, with total cattle numbers including both beef and dairy, in excess of 132 million head in the U.S. After a sharp liquidation phase, there was a momentary pause, and what appeared to be the start of an expansion phase started in the late 1970's. Unlike prior expansion phases, however, this one did not last. It lasted more nearly two to three years than the traditional six to seven years, and then liquidation started again. Numbers were eventually pushed below 100 million head in the late 1980's.

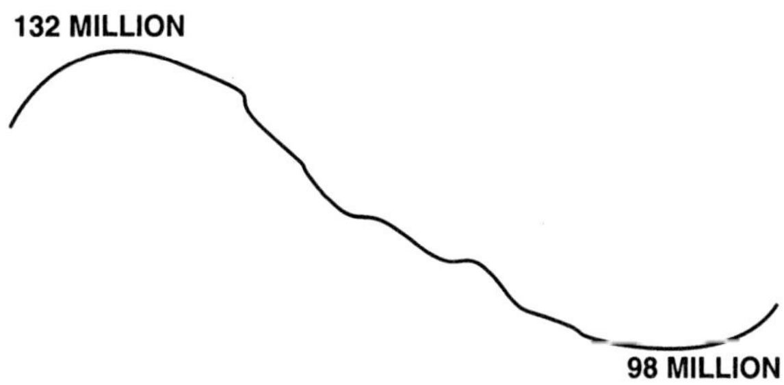

Graph 17 - Major and persistent decline occurred from the late 1970's through the 1980's

Different analysts would offer different reasons why the apparent expansion phase in the late 1970's didn't last. The most plausible and logical reason seems related to what happened to demand for beef starting about 1979. There will be more coverage on this concept in later chapters in the book, but it is important here to suggest that decline in demand started to put downward pressure on price and appears to have aborted the budding cattle expansion.

Beef demand influences the cattle cycle

Graph 18 looks somewhat complicated initially, but it contains some very basic economics and helps us understand why the increases in cattle numbers in the late 1970's didn't last. What was happening was the switch from a supply such as that labeled S back toward one labeled S* as the liquidation phase of the cattle cycle that was going on in the late 1970's ran its course. This is the shift to the left in the supply curve up a negatively sloping demand curve (labeled D) that should have brought us back to higher prices and sustained the expansion motivations in cattle. But this time, something else appears to have happened. The demand for beef at the consumer level shifted down from D to a lower and negatively sloped curve like the one marked D*. Notice that even with supplies reduced to S* on the lower demand curve, supply and demand match each other or balance each other at a price P_I. This is not the high price that was needed to sustain the expansion but is an intermediate price relative to P_L and the potential P_H. In fact, that price suggested by P_I, on occasion in the early 1980's slid back to and below the levels suggested by P_L.

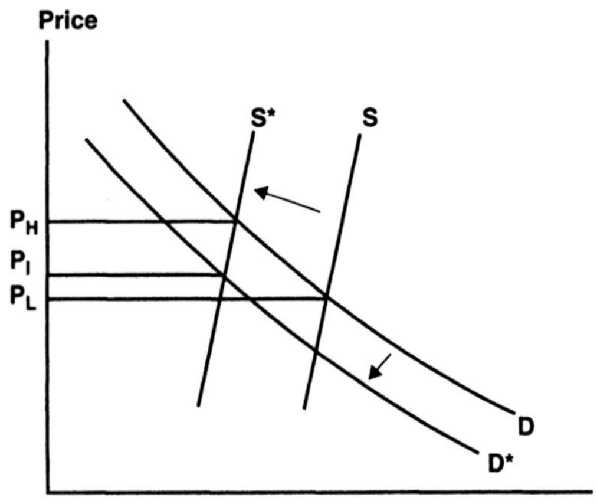

Graph 18 - Decrease in beef demand as a factor in the cattle cycle of the 1970's and 1980's

Granted, all this complicates the cycle in cattle somewhat. But this apparent break from a long history of repetitive cattle cycles doesn't negate the need to think about the cycle. In 1994, for example, numbers have been increasing at a relatively modest level since the "bottom" in 1989-90. The astute cattle producer will start to think about how much longer this developing expansion can be sustained.

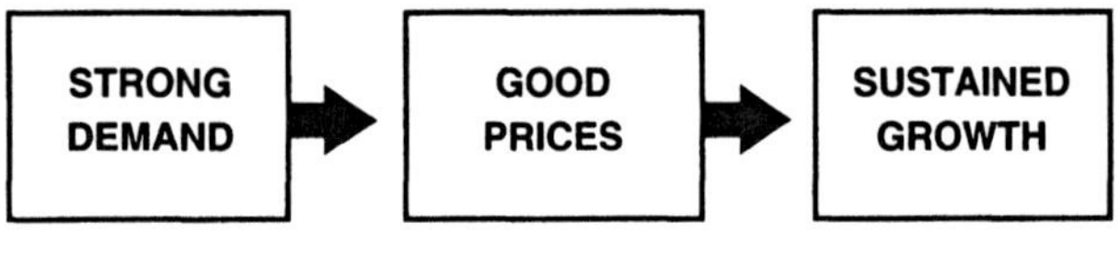

The elements of sustained growth

That line of thought will mean trying to get a handle on whether demand is higher than it was when the expansion aborted in the late 1970's. It also involves trying to get a handle on the rate at which other producers are expanding and avoid getting caught in a "herd mode," where everyone is overexpanding. Just like earlier cycles, overexpansion is sure to bring dramatically lower prices later on.

In simpler terms, even though the 10-year cycle in the cattle business appears to have been broken, this does not mean that monitoring what is occurring at a given point in time is not important. *Any time a producer finds himself facing such an expansion pattern in cattle numbers, it is inevitable that eventually increased numbers of slaughter cattle will come to market and drive prices down.* This is going to happen almost regardless of what happens to demand. So, astute producers will be aware of this and make management decisions to handle their exposure to price risk accordingly.

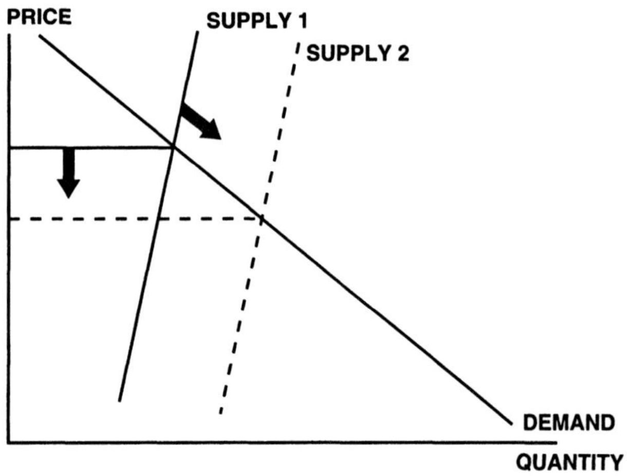

Graph 19 - Higher inventories guarantee lower prices

A cycle in the grain?

In grains and oilseeds and other storable commodities, some people argue that there is no price cycle. That may be right, but there is good reason to reflect on the implications of being caught up in the cyclical-type behavior that is so characteristic of the livestock commodities. *It is very important for the individual producer to develop a perspective that tends to keep them away from the "herd instinct" referred to earlier.* Getting caught up in the flow and going along when everyone else has already started expanding hog production is a sure way to have an increased supply of hogs to sell at a loss. Going along to expand, say, corn production when everyone else is expanding is a sure way to encounter corn prices below your cost of production. So it is important to be aware that the same sort of behavioral dimensions are there in the grains, oilseeds, cotton, peanuts, and other storable commodities as well as in the livestock commodities.

Big increases in acreage will lead to a down price "cycle" in crops

Graph 20 helps build some much-needed perspective. Shown are planted acres in corn, wheat, and soybeans by marketing years since 1960-61. Prior to 1975, most of these marketing years started on October 1 and ended the following September. Since the 1970's, the crop year for all of the summer-grown crops, such as soybeans and corn, start on September 1 and go through the following August 31. The crop year for wheat starts on June 1 and goes through the end of the following May. Thus, when we are talking about planted acres for a corn crop year like 1993-94, we are talking about the acres that were planted for the crop that was harvested in the fall months of 1993 and used for various purposes during the marketing year from September 1, 1993, through August 31, 1994.

GOVERNMENTAL EFFECTS

There are several noteworthy developments in the plots. Perhaps the most significant is the huge surge in planted acreage that occurred in most of the crops starting in and about 1973-74. This surge was associated with a decision in the former USSR to come into the world market and buy feedgrains, foodgrains, and oilseeds instead of liquidating livestock and poultry programs as they had done historically when they had a low crop. Prices of corn jumped toward $4.00 for the first time in U.S. history; wheat prices went sharply higher, reaching the $4.00 level in 1974-75; and soybean prices surged and actually went to double-digit levels periodically in that 1973-75 period. Producers obviously reacted and increased planted acreage in the following years. Some observers would suggest that the acreage response was aided and abetted by the U.S. government. That is true. Production constraints on farmers were relaxed by the U.S. government during that time period, and there was encouragement to plant acreage and rebuild the stocks that had been essentially eliminated by the massive USSR purchases.

It's still basic: high prices will bring more acres

Acreage surges when prices are high

The interesting thing about the data as one moves through it, is the tendency for planted acreage to surge in response to favorable prices and then for acreage to decrease when the prices are not so favorable. We have to always keep in mind that there is a government program in place for wheat and corn. There is usually a set-aside requirement to be eligible for government support, etc., so this does influence the data. The 1983-84 marketing year for corn, for example, is a noteworthy example. In that particular year, because stocks had reached burdensomely large levels in corn, the government announced the much-discussed PIK program, which basically took some entire farms out of production. With acreage pulled sharply lower, the corn sector was then ravaged by drought, and average yields in the U.S. were at 81 bushels per acre that year in sharp contrast to 113 bushels per acre the prior year. Stocks were pulled down dramatically, and the average corn price for the marketing year 1983-84 was $3.25 per bushel. Then the predictable occurred. A sharp increase in planted acreage the following two years generated so much product that, in the 1986-87 crop year, the U.S. average price for cash corn was only $1.50 per bushel!

Big increases in supply lead to huge price declines

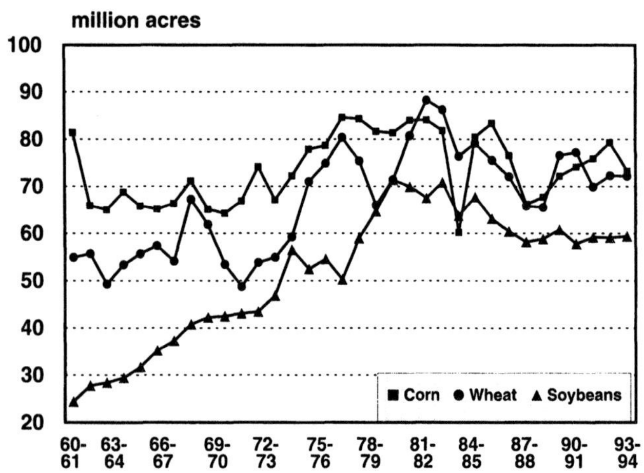

Graph 20 - Planted acres in corn, wheat, and soybeans: 1960-61 to 1993-94 market years

What we see, obviously, when the three key commodities are examined closely, is the same tendency that was evident in the livestock commodities. *Individual producers are inclined to react to favorable prices by taking actions to increase production.* When one farm producer does this, he or she has no significant impact on price. When hundreds of thousands of individual farmers take this type of action, they shift the supply curve dramatically to the right, just as we saw with the livestock commodities, and price is driven lower.

It's what the crowd does that matters

Whether the observer wants to call what happens in the grains and oilseeds a price cycle is not important. The important point here is that a **micro-macro** issue is involved. One way to think about price cycles in all commodities is to think about them in a micro-macro context. That is, the individual producer's actions do nothing to change market price significantly. But when many individual producers take the same action, then price can be impacted in a dramatic way. We will talk about the **micro-macro trap** in later chapters--and it is very important. The individual cannot avoid being hurt by the collective (macro) actions of others.

What is a "micro-macro" trap?

Good advise on the cycle

Perhaps it is best to leave this discussion of price cycles with a warning to the individual decision maker at the farm level. Try to avoid doing what everyone else is doing. There is an old adage that reads, "Run when other people walk, and walk when other people run." That is not bad advice. When everyone else is convinced that hog prices are going to $70 per hundredweight and they are buying gilts and holding back their own gilts to add to their breeding herd, it is probably a good time to start thinking about downsizing your operation a bit or trimming your size for a difficult period ahead in terms of prices. Also, look at what you can do to forward price hogs across the next year or longer. When everybody you see as you drive through the countryside is planting acreage to wheat, then you have to think about what this will mean for wheat prices one and two years later. Is the U.S. or the world market really flexible enough to absorb a big increase in wheat supplies, corn supplies, soybean supplies, oats, cotton, or whatever without a major reduction in price? Chances are the answer is "no." Anything that shifts that supply curve to the right and down a negatively sloping demand curve will reduce prices unless something occurs to shift the demand curve up and to the right.

In the case of the early 1970's experience with the former USSR, the next year the USSR crop was better. Soviets did not enter the market at anything like the same magnitude that had prompted the upward spiraling prices in the U.S. and the resulting surge in acreage.

A rule to remember

All of these markets are charged with risk and uncertainty, and certainly these price cycles are a major player in making prices uncertain to the individual producer. The best advice is to step back and examine whether or not the price increase you are seeing in hogs, corn, or cotton has any reasonable chance of being permanent or whether it is short run. *If you think it is temporary, then it is a mistake to make a long-term investment, such as buying new equipment, adding land, or adding buildings to your operation in response to a short-term incentive.* With that piece of advice, we can leave this topic and come back to it again in the next chapter where more detail is provided and in still later chapters where we deal with price risk management techniques.

SUMMARY

Prices for agricultural commodities move in patterns over time. Within the year, monthly or weekly prices will show a seasonal pattern. For storable commodities, the seasonal moves in price follow a distinctive and predictable pattern: prices tend to be lowest at harvest and then increase during the year reflecting storage costs. Understanding the seasonal price pattern is important to store-sell decisions. Prices will not always increase enough to more than match storage costs, and the storage program can cost the producer money.

For non-storable commodities, the seasonal price pattern reflects the within-year pattern in production. Slaughter hog prices, for example, tend to be lowest in October and November when the pigs born in the mild weather of the spring months—a favorable time for producers—reach slaughter weights and come to market. Milk prices are lowest in the late spring months when lush pastures boost milk production. Beef calf prices are lowest in the fall months when cattle come surging to the market because the pastures are no longer growing.

A seasonal index can be calculated for each commodity. The index value for each month shows how price for that month compares to yearly average price. The index values can be used to predict prices later in the year, an important help to the producer trying to decide whether to sell or hold the cattle or the producer trying to control production to increase sales during the months during the months with the highest prices.

The price cycle is a longer run price pattern that spans several years. It occurs because producers tend to overreact to favorable prices, increase supplies, and drive prices down again. This is a specific example of the micro-macro trap which is referred to throughout this book, and is a prime example of when effective price risk management strategies are needed. In later chapters, as those risk management strategies are developed, the cyclical moves in price will be important.

It is important that the individual producer understand time-related price patterns. The seasonal and/or cyclical swings in price can destroy profits and put the business under financial pressure. But the other side of the coin is also there: the producer who is well informed can take advantage of knowledge of pricing patterns over time. How that can be done will be covered in each of the following chapters.

3

The Deadly Micro-Macro Trap in Production Agriculture

INTRODUCTION

In Chapters 1 and 2, there was brief reference to the micro-macro issues facing the individual farmer. There are not many things in the environment in which the individual grain, livestock, fiber, or vegetable producer operates that are more important than this issue. We come back to it here in Chapter 3 with a focus on this "trap" and why it is so important. There is not much we can do to develop marketing strategies that will work for the producer unless the potentially devastating implications of this micro-macro problem are understood and handled.

THE MICRO-MACRO SETTING

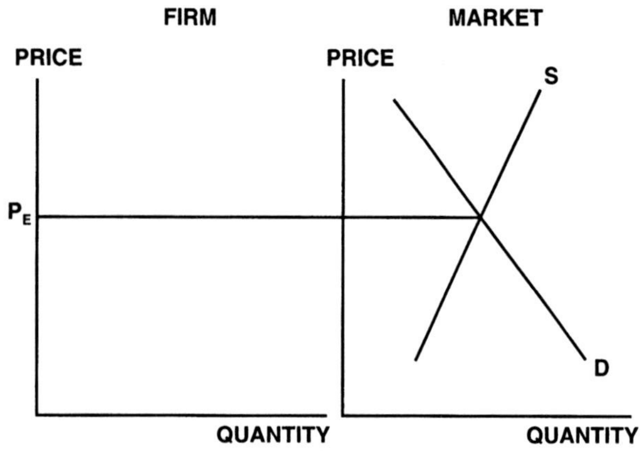

Graph 1 - The market-wide price facing the individual farmer

Graph 1 shows the important producer-market relationship again. At the market level, interaction of demand and supply determines a price. The individual producer faces this price, the price labeled PE in the figure, and is a price taker. Chapter 1 introduced the idea of why prices move over time, and Chapter 2 stressed that movement across time in some detail. The emphasis in Chapter 2 was on shifts in the market-level supply curve brought on either by seasonal or cyclical patterns in production. We found that the dominant factor that moves price within the year and from year to year is shifts in supply of agricultural commodities. That having been said, the important thing for the individual producer is to avoid the traps that loom for the unsuspecting.

You have to avoid the "trap"

Let's think about what happens when something occurs at the market level to push price up. It is possible that the cause is increased demand at the world level spilling into our markets because some major producing country has lost part of its crop to weather problems. This was the primary issue behind the much-discussed USSR purchases in the early 1970's. Alternatively, it might be weather problems for crops in the U.S. or in some other major country that push price up. Whatever the source of the price increase, let's imagine a price facing the farm that is above the price labeled PE. What does this mean? How is the manager at the farm level likely to react?

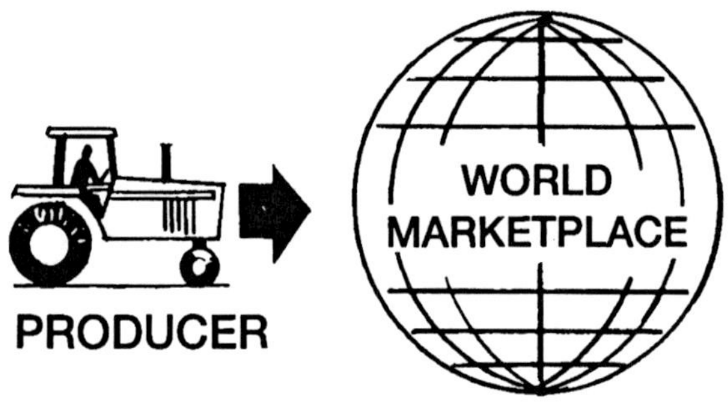

The producer has to react to world-wide developments

To explore this, let's assume that there has been a shift in the supply curve to the left. A reduction in supply pushes the price up from PE to a price we call PT. This is shown in Graph 2, and note that the PT price is well above the original price.

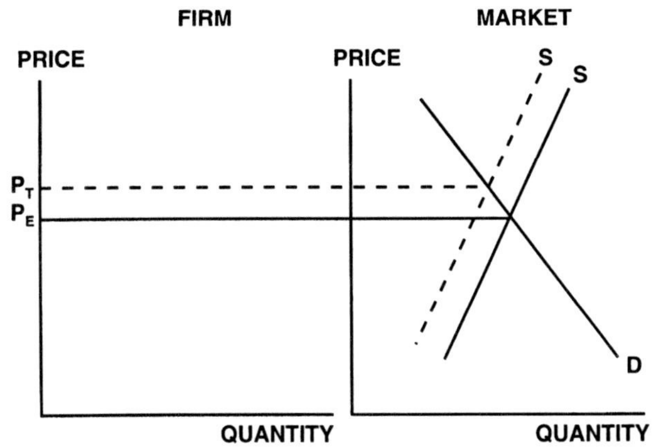

Graph 2 - The price implications of a decrease in supply due to weather problems

Readers with some background in economics will know that the price facing the individual firm is also its marginal revenue. In other words, the change in revenue associated with producing and selling one more unit of, say, corn is the price at which you can sell another bushel of corn. There are books written about marginal analysis in basic economics. To maximize profits, the producer should operate at the quantity where marginal revenue equals marginal cost, or MR=MC. *The individual producer will want to expand output so long as the change in revenue associated with another unit of output exceeds the change in cost.* If we can get back $3.00 for an added cost of $2.50, profits will go up. To maximize profits, output should be expanded until that increment of MR above MC is pushed down toward zero.

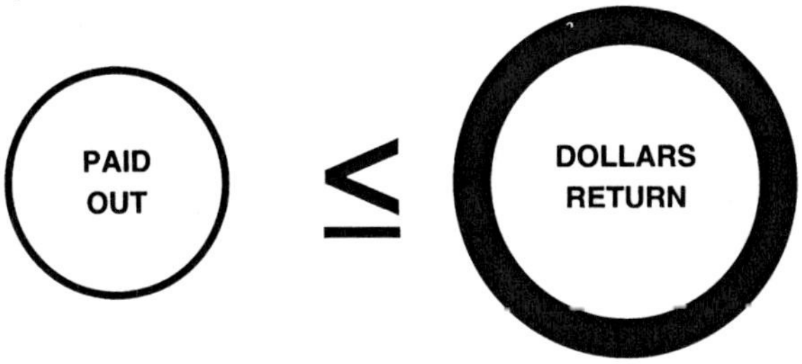

Take an action only so long as what you get back exceeds your cost

It will go to zero eventually. When we push that production process harder, the additional cost associated with each added unit of output, in this case each added bushel of corn, starts to go up. Eventually, at any particular price being discovered in the marketplace and facing the individual farm, output levels can be adjusted such that the flow coming back to the producer (MR) is just equal to the flow out (MC). At that point, the individual farm is maximizing profits.

MARGINAL REVENUE IS TEMPORARILY ABOVE MARGINAL COSTS

Don't make this too complicated. Think about what is going on in the minds of the decision maker after the price has jumped up to PT as shown in Graph. 2. Now, assuming the costs of producing this corn have not changed significantly just because the selling price has gone up, the individual producer faces what looks like a pleasant situation. Corn price is up to some very favorable level, such as $3.25, if we assume the PT price has zoomed to that quite high level. This individual farmer can produce another bushel of corn for less than $3.25, perhaps $2.25. In other words, the **marginal revenue** now exceeds the **marginal cost** ($3.25 is greater than $2.25), and every beginning economics book in the world will tell you that, in those circumstances, the producer should expand output and produce more corn. But at this point, the micro-macro trap rears its ugly head. *It is very important that the individual producer be able to step back and recognize this situation for what it truly is.*

The micro-macro trap is on the horizon

Two ways for the producer to react

There are two ways the individual producer can react to this apparently attractive price. One of them is the correct approach; the other is decidedly the wrong or incorrect approach. What we need is some effective guidelines to allow the producer to make the right decision and respond to this attractive-looking price correctly.

The first thing we have to decide is whether or not this increase in price is likely to be temporary. This decision should not be as difficult as it may sound on the surface. If corn prices are higher because a drought in the Midwest wiped out part of the corn crop last summer, then as the individual farmer looks at a relatively high corn price and is thinking about how many acres to plant in May of the following year, this ought to clearly be seen as a temporary situation. *There is no reason to argue that just because there was a drought last year, there will be a drought this year to hold prices up at the, say, $3.25 level that we have assumed for illustrative purposes.* Weather in one year ought to be relatively independent of weather in another year, and if last year the 1-in-20-chance drought was actually experienced, the odds of having a drought again the current year are still about 1 in 20. There is nothing to suggest that weather disruptions for a crop, whether it be corn, wheat, cotton, peanuts, hay, vegetables, or anything else we are producing, is anything other than a temporary phenomenon.

Is the high price temporary?

In many instances, any price increase that is brought on by an increase in demand in the world market can also be temporary. This was true of the already-discussed purchases by the former USSR in the early 1970's. The USSR came into the market and bought grains to offset a short crop to keep from having to liquidate poultry and livestock. The next year, when weather was more nearly normal and the crop in the USSR returned to normal, demand in the world market shifted back down. *The increase in demand was temporary.* The stratospheric prices that we had seen for both feedgrain and foodgrain crops in the U.S. started to disappear.

Weather problems are always temporary

Or is the high price perhaps permanent?

If the shift up in demand has some reason to be permanent, then one might anticipate the price strength would remain. Let's assume, for example, that legislation is passed in the U.S. that requires X percent of every gallon of gasoline to be made up of ethanol, which is an alcohol produced from corn. This is not out of the realm of possibility given the needs to reduce air pollution, etc., and if such legislation were passed, that might push corn prices up substantially with some very real possibility of them staying high over a number of years. But the important point here is that the individual producer has to step back from this scenario and decide whether or not the price increase has any reason to be considered permanent or whether it is more likely to be temporary.

If the decision is that the price increase is likely to be temporary, then the correct response is more *variable resources* to increase production. You use more labor, more fertilizer, more machine time, and do the other things that you can to increase production of corn in response to the higher price. *But all of these adjustments are flexible.* Say you put another 20 lbs. of nitrogen per acre on the crop this year because price is higher, if price is back down again next year, you can reduce that level of nitrogen back to long-term levels. If you hired someone part-time this year to help you cultivate and handle some additional acres, given the higher price, then next year if price is back down, you don't hire that part-time laborer. *In other words, when you increase output by using variable resources, you leave yourself in a variable and controllable situation.* Any decision that is made that increases output of corn on your farm is a reversible decision, and you don't have to continue to operate at that high level of production next year.

Using more variable input is a reversible decision

The big problem

The problem occurs when many farmers make the incorrect response to a temporary price increase. By incorrect, we mean bringing in more *fixed resources* to expand production. This means that instead of just using more fertilizer, hiring some part-time labor, and using your tractor more hours, you buy new equipment, more land, or put up additional buildings and storage facilities. These adjustments to the higher price may allow you to increase output more, but they also carry with them a very important characteristic: *Buying land, buying more equipment, and putting up new buildings is not a flexible, reversible decision in the short run.* If prices of corn are back down next year, it is not easy to reverse the processes that caused you to expand corn output this year. You are stuck and these resources are "trapped" in production. If you bought another tractor or another combine to allow you to handle more acres of corn, you essentially have to keep that equipment and use it. The salvage value for used equipment is often, in these circumstances, below it's value in use. Other farmers who got caught making the same mistake also have equipment they would like to sell. Typically, these resources are kept in use.

Land, buildings, and equipment increase output, and these investments are not easily reversed

It is important to recognize that if a number of farmers make the incorrect response to a temporary price increase, then a price decrease is guaranteed. Using more variable resources, such as labor and fertilizer does not shift the industry-level supply curve out and to the right, i.e., it does not increase industry supply. Bringing in new capital equipment, such as more land, more buildings, and new combines and tractors, *does* shift the industry supply curve to the right.

What we have, then, instead of the industry supply curve being at S′, it may, after this incorrect reaction by a number of farmers, be at the original level of S or even to the right of S. This suggests that industry supply is now greater than it was before the price stimulus occurred. In fact, that is exactly what we would expect to happen given that new equipment, new land, and new fixed resources have been brought into production. *This is the micro-macro trap. Making a long-term and irreversible response to a temporary price increase can be, and typically will be, devastating to the individual farmer.*

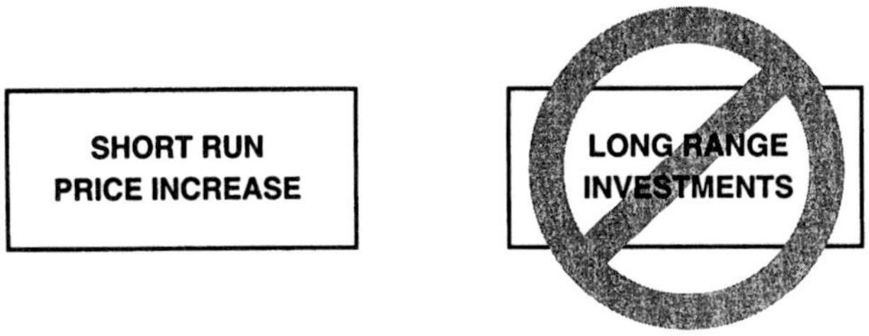

Don't respond with a long range investment

DOUBLE TROUBLE

What happens after these new capital expenditures were incurred is that price is driven back down. The costs for these farmers are now higher than for their neighbors who did not make such an adjustment. Any cost-price squeeze as corn prices are pushed back down toward $2.00, for example, is harder on the producer who bought the new resources than the one who didn't. With rare exception, and that would occur when economies of size from a big operation actually pull overall costs down, costs per bushel of producing corn will be higher after the new equipment and new land are brought into production than they were before. The producer who fell into the trap is then doubly in trouble. Those very actions, when spread across many thousands of producers, force the industry supply curve out to the right and force price back down.

Precisely this type of adjustment to initially higher prices characterized the agricultural sector in the U.S. during the 1970's and into the 1980's. New land, new equipment, and new fixed resources were, in fact, brought into corn, wheat, and soybean production. Prices were driven sharply lower, below cost of production for producers who now had inflated costs. The farm crisis of the 1980's, as many farmers were forced to declare bankruptcy and farms were sold to pay off creditors, was the lingering aftermath of this massive set of bad decisions in the 1970's. It is only now, in the mid-1990's, that the level of farm debt is going back down toward historical norms. 20 U.S. grain and oilseed markets are starting to settle down some 20 years after the initial shock of those infamous USSR purchases.

Land was put back into production after U.S.S.R purchases in 1973

The micro-macro trap can destroy

The micro-macro trap can destroy the individual producer. The already-mentioned adage of "walk when other people run and run when other people walk" fits here. The ability to back away from the ecstatic talk at the coffee shop, when corn prices are pushed up to $3.25 per bushel, is essential. Some breadth in perspective and some ability to recognize that when you as an individual feel the urge to buy new equipment, buy new land, and expand, that same urge is being felt by thousands of other producers. *Whether you individually do so will not matter, but whether thousands of other producers expand will matter. The aggregate or macro actions of a group of producers can and will drive prices sharply lower.*

If everyone increases production, prices will fall

Later, as we work on marketing strategies and talk about marketing management, we will come back to this point and make sure that this decision is discussed again in a proper context. Immediately, it becomes important that if you are going to expand output in any way, even with increased usage of variable resources, you need to recognize that others are likely to over expand and push price back down. That raises the need to get the expanded production that you planned in response to the drought-induced higher prices last year forward priced early in the year while prices are still favorable. But we have to recognize that it will be unlikely that you can always, as an individual, escape the pain of the macro response from thousands of other producers around the country. Corn, soybean, and wheat prices are discovered on the Chicago Board of Trade (CBOT), and the futures market there is in the business of discovering price for future time periods. In the year following a drought-reduced crop, the futures market traders understand that we are likely to see expanded acreage to shift the supply curve out and to the right and to drive price back down. Thus, as soon as it becomes apparent that producers are starting to think that way, the prices being discovered for some future month at the CBOT will start to adjust downward. So, it is especially important that the astute producer be out front on this issue and get some forward pricing done early.

The surge in the average is the problem

To emphasize the importance of this issue, Table 1 shows again the acreage planted to corn starting in the early 1970's and more recent years. The surge in new acreage being brought into corn production extended into the late 1970's. It was this surge that brought fragile land into production. Federal government programs designed to force some of this land back out of production have been established in recent farm legislation. But the point is that it was very difficult to reverse that decision to plow up grassland in the Southwest, plant additional acreage in the Midwest, or expand production by draining wetlands wherever they might have been in the U.S. That acreage stayed high and led to disastrous corn prices that extended into the 1980's. *This table needs to be kept in mind as the reader tries to form a lasting impression of the economic pain that can be associated with making the wrong decision in response to a price stimulus that could turn out to be temporary.*

The micro-macro trap in livestock

The same trap is there in livestock. In Chapter 2, there was much discussion about the swine sector in the U.S. and how periodic over-production and over-expansion pushes that industry supply curve out and drives prices below cost of production. It happens more frequently and occurs more often than in corn. Therefore, for the livestock producer, this micro-macro trap is in some respects more important.

Year	Acres Planted (mil. acres)	October Price ($/bu.)	Year	Acres Planted (mil. acres)	October Price ($/bu.)
1975	78.719	2.62	1985	83.398	2.11
1976	84.588	2.33	1986	76.58	1.40
1977	84.328	1.67	1987	66.2	1.55
1978	81.675	1.97	1988	67.717	2.58
1979	81.394	2.41	1989	72.221	2.22
1980	84.043	2.99	1990	74.171	2.19
1981	84.097	2.45	1991	75.951	2.31
1982	81.857	1.98	1992	79.3	2.05
1983	60.207	3.15	1993	73.3	2.50
1984	80.517	2.65	1994	79.1	2.15

Table 1 - planted acres in corn and october corn prices 1975 to 1994

Year	June-November Sows Farrowing (1,000 Head)	% Change from Previous Year	Price of Hogs Next Year
1975	4,952	-9.569	43.83
1976	5,850	18.134	41.30
1977	6,009	2.718	48.46
1978	6,398	6.474	42.48
1979	7,322	14.442	40.04
1980	6,855	-6.378	44.45
1981	6,268	-8.563	55.44
1982	5,884	-6.126	47.71
1983	6,181	5.048	48.86
1984	5,857	-5.242	44.77
1985	5,669	-3.210	51.19
1986	5,440	-4.040	51.69
1987	5,810	6.802	43.39
1988	6,036	3.890	44.03
1989	5,775	-4.324	54.45
1990	5,717	-1.004	48.88
1991	6,074	6.245	42.31
1992	6,006	-1.120	43.38
1993	5,947	-0.982	39.63
1994	6,223	4.641	

Table 2 - U.S. sows farrowing, percent change in sows farrowing from previous year, yearly average slaughter hog prices, 1975-1994

To emphasize the difficulties in livestock, Table 2 shows farrowings in the December 1 Hogs and Pigs Reports since 1975. These farrowings in December 1 of any particular calendar year are matched or aligned with the average prices for slaughter hogs for the following calendar year. By showing farrowings as percentage changes compared to year-earlier levels and the prices that result from those periodic sharp increases, it is apparent what happens when producers get eager and expand too much. It bears repeating that the producer who makes an investment in expensive new facilities is going to be hurt the most when prices are driven below $30 per hundredweight within the year, less than anybody's cost of production.

The moral to this story, focusing on the swine sector, is if you are going to make long-term adjustments and long-term investments in new buildings and facilities, make sure you understand what type of price pattern you are likely to see over time. Plan to be able to operate with periodic prices below cost of production, and plan to do a good job of forward pricing to protect yourself against the cyclical reductions in price. There is no question that in hogs, it is the periodic and cyclical-related movement of the supply curve that drives the prices up and down.

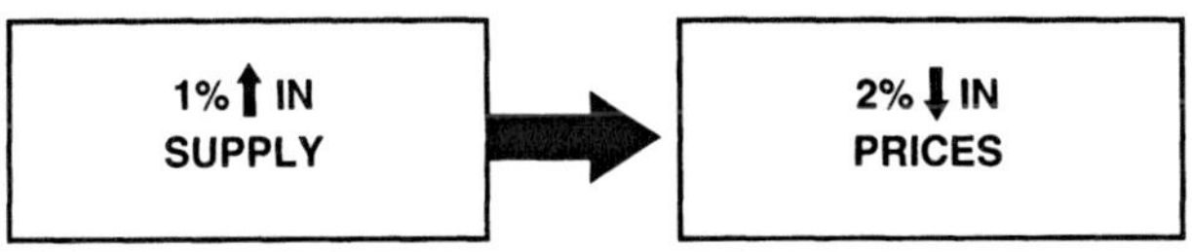

A rule of thumb is a price swing double the supply change

The incorrect response is the problem

Keep in mind, however that the primary thing that shifts the supply curve to the right, overexpands the industry, and drives prices down sharply is producers making the incorrect response to a temporary price increase. New farrowing buildings are built, new feeding floors are brought in, new facilities are put in place to expand output, and eventually that expanded output comes to market and drives prices back down. Then, those producers with the new facilities are in major trouble. If you are going to make a long-term investment in swine production, for example, make sure you can handle the periodic swings in price and make sure you are an efficient producer and will be able to survive by using good production management and good marketing management over time.

As the "trap" appears, it's time to think about forward pricing

SUMMARY

The micro-macro trap demands special attention. Making the incorrect response to a temporary price increase has ruined many producers. If a price increase is likely to be temporary, then the producer should just respond by using more variable resources--and staying flexible. Keep in mind:

- **You can't stop other producers from making a mistake.**
- **Overly zealous responses to better prices will, if new capital is brought in, shift industry supply to the right.**
- **An increase in industry supply will push prices down--and your prices will fall too.**
- **The producer without the added capital investment will be better prepared to withstand a period of low prices.**
- **As the "trap" starts to appear, it is time to think about getting prices established.**

This last point moves us toward forward pricing again. Chapter 4 introduces futures markets, the reason we can forward price.

4

Trade in Futures and Options

INTRODUCTION

At many points in this book, the agricultural producer is described as a price taker. This means he or she has no capacity to influence selling price for the agricultural product. There are many relatively small producers. No one is large enough to influence the market-determined price on any particular day. This is true of wheat, slaughter cattle, cotton, and it is true of many agricultural commodities that are broadly produced in the United States by a large number of quite small producers.

The market place is bigger than the individual producer

PRICE RISK MANAGEMENT

Because the individual producer is a price taker, we have found that it is very important that they manage their exposure to the variable prices effectively. In other words, *price risk management* programs must be put into place and they must be effective. Managing exposure to price risk requires an understanding of and a willingness to get involved in forward pricing. But forward pricing would not be possible if there was no trade in futures contracts of agricultural commodities. The buyer who might be willing to establish a price for later delivery with an individual farmer has to have some place to manage his own risk exposure. As we look at this very important need, we will find that it is trade in futures contracts that provides the mechanism that allows for risk transfer.

What we have is an organized effort in a number of relatively large exchanges around the U.S. to discover a price for agricultural commodities for some later time period. Many, but not all, of the agricultural commodities are traded on the two large exchanges in Chicago. The Chicago Board of Trade (CBOT) offers contracts in wheat, corn, soybeans, soybean meal, soy oil, and several other agricultural or agricultural related commodities. The Chicago Mercantile Exchange (CME) offers contracts in slaughter cattle, slaughter hogs, feeder cattle (cattle of the size and weight ready to go into the feedlot), pork bellies, lumber, and some other commodities that have implications to agriculture. On exchange in New York are futures contracts for cotton and for orange juice. All of these exchanges offer futures contracts in interest rates, stock indices, petroleum products, and a number of other commodities and services. A futures contract for milk is being considered by the CME. Generally, what we are finding is that any time there is substantial and broad exposure to price risk, we have an exchange that is willing to offer a futures contract for that particular commodity to allow cash firms to manage their exposure to risk.

FUTURE DELIVERY

The price discovery process is complex, and need not be covered in great detail here. The important thing to recognize is that contracts for delivery as much as one to two years in the future are traded on the futures exchanges. This means that at any particular point in time, buyers and sellers are trading contracts calling for delivery of agricultural commodities as much as two years in the future.

By forward pricing, at least I can control when price is established

FUTURES TRADE AND CASH CONTRACTS

This price discovery process is offering a major service to the individual farmer. We will find that because there is trade in futures and options on futures, the farmer *does* have an opportunity to monitor the prices being offered and to exercise *control over when price is established*. As has been noted earlier, about the only thing the individual producer can control is when that price is established. Trade in futures makes it possible for buyers to offer cash contracts. Grain elevators have offered cash forward contracts for grains and oilseeds for many decades. The forward contracts for slaughter livestock have grown substantially in popularity in the 1980's and 1990's. It is useful to pause before we look at the process in more detail to recognize that *there would be no possibility of cash forward contracts if there were no trade in futures.*

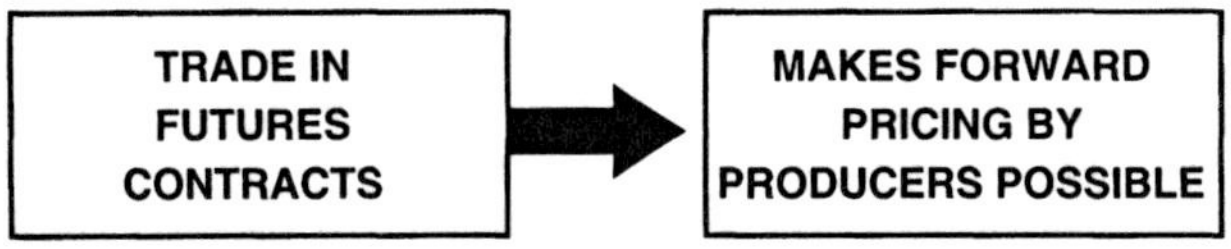

We must have the futures markets

Intuitively, then, it appears that the cash contract is tied directly to the futures and it is. No grain elevator would be willing to offer anything but a very low price for corn or cattle to be delivered later if the elevator had to absorb the pain prices decline sharply. The futures markets allow the elevator to transfer that risk--and we will see it all works out to the benefit of the farming sector.

WAYS TO FORWARD PRICE

There are three basic ways to forward price that will be explained and demonstrated very briefly in this chapter. The first is the **cash contract** that calls for later delivery of a commodity, an approach that has been around for a long time. As suggested in the introductory comments, buyers of agricultural commodities can offer these contracts because there is trade in futures that allows them to transfer their own risk exposure to others. The second approach to forward pricing involves **using the futures directly**. What this means is that the individual producer takes a position in the futures market to establish a price for some time in the future instead of allowing the grain elevator or the meat packing firm that is offering cash contracts to do it for them. The third approach to forward pricing is much newer than the others and really has come into popular use since the mid-1980's. This approach uses **options on a futures contract** which offers an intriguing additional possibility to the producer. It offers a price floor instead of a specific price as would be the case in a cash contract or direct trading of futures. In this chapter, we will look at each of these approaches to forward pricing briefly and establish a framework use in later chapters as we deal with approaches to price risk management.

The cash contract has long been familiar in commodity marketing

CASH CONTRACT

Any cash contract for an agricultural commodity tends to have some very basic provisions:

- **The contract calls for later delivery of a commodity of a specific type and quality.**
- **The contract price is tied directly to the price being discovered in the futures markets.**
- **When the producer signs a forward price establishes price in a cash contract, his exposure to price risk has been transferred to someone else who is trading in the futures complex.**

Cash contracts can call for later delivery ranging anywhere from a day or a few days up to as much as two years. How far into the future is less important, perhaps, than the fact that the buyer and seller for their own particular reasons want to get price established. Buyers of wheat, for example, might know that they are going to need a substantial volume of wheat to meet an export commitment later in the year. They might want to make sure that they have adequate supplies to meet that commitment. Therefore, they may be very interested in negotiating a cash contract with the farmer that calls for delivery, say, at harvest time or possibly sometime after, assuming the farmer can store the commodity. The farmer, in turn, might be very interested in getting that price established because he has a note to pay at the bank, must meet living expenses, and has expenses with regard to running the farm. Having a price established may make it much easier to move on through the production period. Farmers will come at this with differing risk aversion levels, but most producers would like to get an attractive price established so they don't have to worry about the price falling and putting them in a difficult financial position.

Living expenses can prompt interest in getting price set

Cash contract tied to futures

The cash contract is always tied to the futures. The bid price for corn, for wheat, or for slaughter hogs that is offered by the buying processor on any particular day is tied directly back to the futures market. What we will find as we move on into demonstration of this process is that the buyer will establish the negotiated price in the cash contract with the producer and then turn immediately and take a position in the futures to eliminate his exposure to risk. Assume, for example, that a grain elevator has signed a contract for later delivery of 10,000 bushels of corn at $2.75 per bushel. Clearly, that elevator does not want to pay $2.75 for corn which could turn out to be worth $2.25 at the time its delivered later in the year. A huge crop has increased supply and driven prices down. When the elevators sign the contracts at $2.75, they turn immediately and sell a corresponding number of bushels in the futures market. This is a traditional "short hedge" that protects the elevator operator in the event that all prices go down and the cash market moves down to $2.25. We will come back and demonstrate procedure on this as we go through the chapter, but it is important to recognize that *the cash contract is offered and is available only because there is an underlying futures market.*

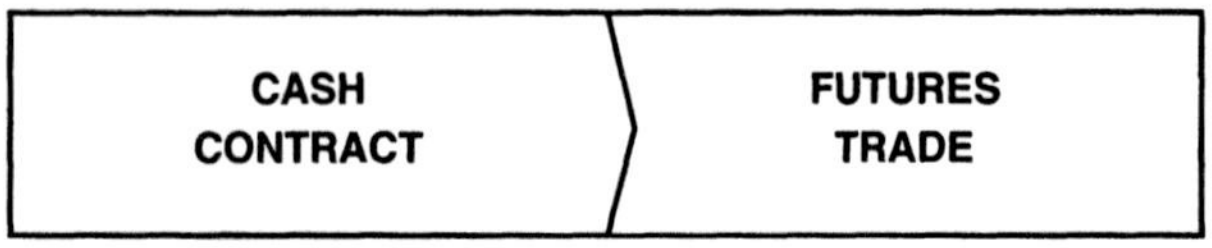

You can't have cash contracts without the futures markets

The third point that is characteristic of the cash contract is that the producer no longer faces price risk. He faces production risk. There may be some concern that since he contracted to deliver 10,000, he may not be able to produce that many bushels. But he does have price pegged at $2.75 (in the above example) so he therefore faces no price risk. Actually, the buying elevator doesn't face a price risk here either, assuming that they did, in fact, cover their position in the futures market. We will find later that the remaining risk exposure here is what we call basis risk. Even that risk has been transferred to the buying grain elevator, and the producer, now free of price worries, can concentrate on doing a good job of production.

Price risk is eliminated by the cash contract

THE TRADITIONAL SHORT HEDGE

The producer can go directly to the futures market and establish a forward price. He does this by selling the futures or, as the trade tends to say, going "short" in the futures market. We will find that certain characteristics of the short hedge will be important:

- **There is always a price being discovered in the futures market that sets the forward price the producer is being offered.**
- **The basis, or the difference between the cash and futures market, is always present and is used to pull the Chicago, New York, or other major exchange-based price down to fit the producer's local conditions.**
- **The effectiveness of the short hedge will always depend on the performance of the basis.**
- **There are some costs associated with the hedge in the form of brokerage fees and interest on margin money that is tied up during the process.**

These concepts will be introduced and explained in this section as we illustrate the basic short hedge.

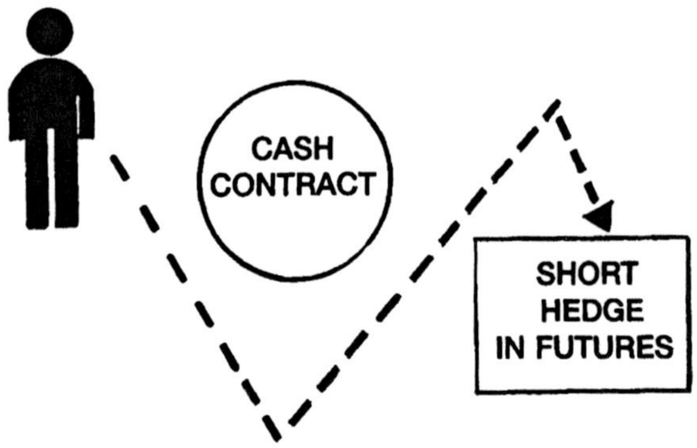

The producer can bypass the cash contract and go directly to the futures market

Two markets discover price

The essence of this entire process is that there are two markets that are discovering a price for the same commodity. If we are talking about a corn producer, there is a cash market that is available at the local grain elevator. This is the traditional market that has always been there for the farmer. But when we add the futures market, we find another related but separate market that is also discovering price for corn. Forward pricing in the futures market directly, called hedging, protects against the declining market because these two related but separate markets will always move together. *This is very important.* To illustrate, if the cash market goes down over time as we move toward the maturity date of a particular futures contract, we will find that the futures market will go down as well. It becomes the basis for a forward price or hedge in the futures market. If the cash market goes down, this decline in potential selling price shouldn't hurt the hedged producer because the futures market will go down with the cash. By taking a sell or "short" position in the futures, the gains in the futures side of the transaction offset the losses in the cash market.

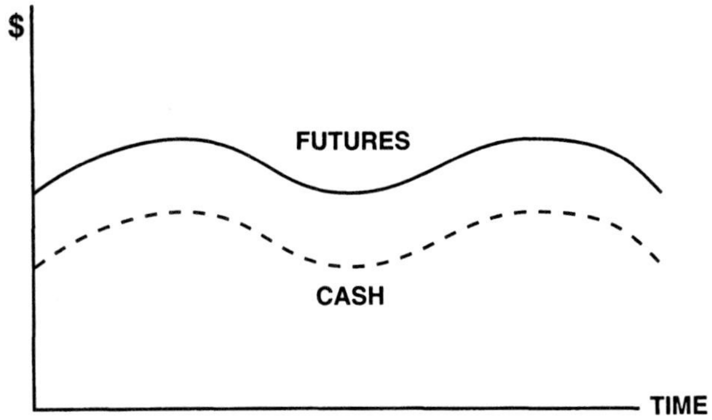

Graph 1 - The key is that futures and cash prices move together over time

Calculating the forward price

To demonstrate this process it is useful to set up a very simple formula that shows what price is being offered by the futures market. The formula is as follows:

FP = FUT + BASIS

where

FP = forward price,

FUT = trading level of the relevant futures con... tract, and

BASIS = cash-futures basis (defined as cash minus futures).

This says, simply, that the forward price being offered by the futures complex is the futures price that is being discovered, adjusted for basis to bring it down to the farmer's local market.

That futures quote is easy to find. It is available in most major newspapers, always available in the *Wall Street Journal*, and increasingly is coming into farmers' homes or offices via an electronic news wire. There are a number of services available to the individual farmer that, after fairly nominal initial set-up fees, bring a stream of price information from the futures complex into the office for as little as $25 to $50 per month. The information is becoming widely available, and there are services clamoring for the opportunity to provide you with this type of price information.

Futures quotes are readily available

Getting ready

Let's assume, to start building an illustration, that it is April and our individual farmer is getting ready to plant corn. He is getting increasingly worried that acreage planted to corn will be up this year because price last year was fairly good. He is afraid that if weather is normal and we have a huge crop, price will be driven down sharply in the fall months when harvest is being completed, primarily in October and November. He starts monitoring the December corn futures, which is the new-crop contract for virtually every producer in the U.S. A quick look at a newspaper or the *Wall Street Journal* will indicate that futures are not traded for every month. They are traded for September and then not again until December, and the December futures are widely used as the month in which a growing corn crop is forward priced. Let's assume the electronic service is showing, on April 10, that the December corn futures are trading around $2.90.

Discovered price can change

What this means is that given the available information on April 10, the buyers and sellers of the December corn futures think that corn price in the Midwest, in the Chicago area, will be around $2.90 come October and early November. This discovered price can and will change as the year goes on reflecting an influx of new information. Particular emphasis is always paid to what kind of weather patterns prevail and how big the crop is projected to be. But on April 10, given the available information, the price being discovered is around $2.90.

CHICAGO FUTURES QUOTE − **CASH FUTURES BASIS** = **LOCALIZED FORWARD PRICE**

The basis adjustment pulls the futures price down to a local market

That price is not really a price that is appropriate for our producer who is located in the southern part of Illinois. What the producer needs is an understanding of what, exactly, a $2.90 price in the Chicago area means to his farming community price in southern Illinois. This is where the concept of basis comes in. Basis has been defined as cash minus futures, and since corn is generally considered to be worth somewhat more in Chicago, where it may be available for processing or going out via river barges or the Great Lakes into export, the futures prices are likely to be above the cash price in southern Illinois. If that is the case, subtracting futures from cash means that the basis will take on a negative level. On any day, the farmer could call the local grain elevator and get an indication of the cash market and compare it to any one of the futures quotes that is being offered from Chicago and calculate a basis. But the basis the farmer needs on April 10 is a very particular one. *What the farmer needs on April 10 is an expectation as to what the basis will be in October when he expects to harvest his crop.* More specifically, he needs to know what, come the next October, the cash price is likely to be relative to the trading level of the December futures.

It is the harvest period basis which the producer needs

Need for basis information

Here then is a need for historical information on basis in the farmer's particular community. It is sometimes available from the local extension agent or from the university specialist. If it has not already been built and made available by someone, the farmer needs to make sure he starts accumulating this information. What is needed is at least five years of historical data to show how the cash price during October has compared to the December futures price during October. Let's assume we have that historical data base in place and, although there is some variability across time, the basis averages -$.10 per bushel. With this information, our producer is now ready to look at what forward price is being offered by the futures complex when the December corn futures are trading at $2.90.

A record of basis levels is very important

We can plug numbers into the formula and calculate the forward price as follows:

$$\begin{aligned} FP &= \$2.90 + (-\$.10) \\ &= \$2.90 - \$.10 \\ &= \$2.80 \end{aligned}$$

For this particular farmer with a historical cash futures basis of -$.10 during the harvest period, a $2.90 trading level in the futures means a forward price of $2.80 is being offered. If the price being discovered in the futures drops to $2.60, for example, then the forward price being offered drops to $2.50, etc.

The reader is probably very interested here if this is a relatively new concept. How does a trading level in the December corn futures of $2.90 offer the farmer in southern Illinois a forward price of $2.80? Where does the money come from? How does the producer, in fact, realize a $2.80 forward price if he goes directly into the futures and takes a position instead of dealing with a local grain elevator? An illustration is needed here, and Table I provides that illustration.

TO SELL FUTURES

Selling futures is just a contractual obligation

The net price

In the table, it is assumed that the farmer sells December

futures on April 10 at $2.90. Then, on October 14, the best price he can get for the corn when he gets ready to sell it is $2.10. This would be a real problem if the farmer had not forward priced the corn. The $2.10 price is at or below the cost of production for many producers. But if the cash market has moved down to $2.10, then we would expect the December futures to move down to approximately $2.20. Remember, our historical database indicates that cash corn is typically, during October, around $.10 per bushel below the trading level of the December futures. Here, we assume that the $.10 differential that we expected when we set up the forward price is in fact exactly realized, and the December futures can be bought back at $2.20. The net price is then the cash price that you receive plus the net from the futures transactions, and we have $2.10 + $.70 = $2.80. The reader must understand this before proceeding.

Date	Cash	Futures
April 10	Will grow corn for sale in October FP = FUT + BASIS = $2.90 - $.10 = $2.80	December futures @ $2.90
April 10		Sell December futures @ $2.90
October 14	Sell cash corn at $2.10	Buyback December futures at $2.20
Net Price	= Cash Price + Net From Futures = $2.10 + $.70 = $2.80	

Table I - Demonstration of a short hedge for a southern Illinois corn farmer

The $2.10 is rather obvious. That is the selling price of the cash corn. The reader might have more difficulty seeing exactly where the $.70 from the futures complex comes from. If you can sell the futures at $2.90 and then buy them back later at harvest time at $2.20, that is a profit of $.70 per bushel in the futures account that is now credited to the producer. He combines that $.70 with the $2.10 he received from selling the corn and is back to the $2.80 forward price.

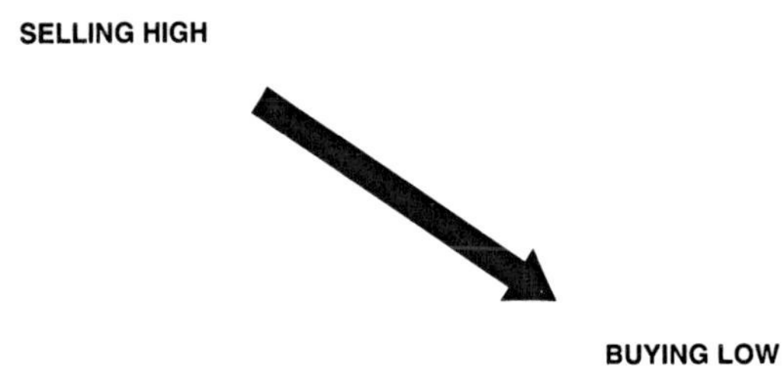

If prices do fall, buying futures back low covers the losses in the cash market

There are several things that still need some elaboration. We will demonstrate hedging and use of the futures market many times in other chapters in the book, but the beginning reader is going to be interested in knowing why you can sell a December futures that you don't have back on April 10 and why, in fact, you buy it back later in the year. Somehow, at first glance, this doesn't make a lot of sense. *But we need to keep in mind that the reason the futures market exists is to allow someone in the cash business such as our southern Illinois farmer to transfer price risk to someone else*. The idea is to take a position in futures, such as selling December futures at $2.90, and then buy back that commitment later.

When those futures contracts are sold on April 10, the farmer does not get any money. What he has really done is accepted a binding contractual commitment to deliver corn to designated midwestern delivery points on or before a final trading date in late December. If he does not buy back that commitment he will, in fact, be legally required to deliver the corn! But the futures market is not set up as a delivery mechanism. It is intended for the producer to sell futures to get price protection and then buy back the futures positions later to cancel that initial commitment. As we can see in this example, it works beautifully. The farmer would be hurt by the declining cash price. But since he or she has been willing to establish price back on April 10 for later delivery, they end up with $2.80, and the operation is in good shape and profitable.

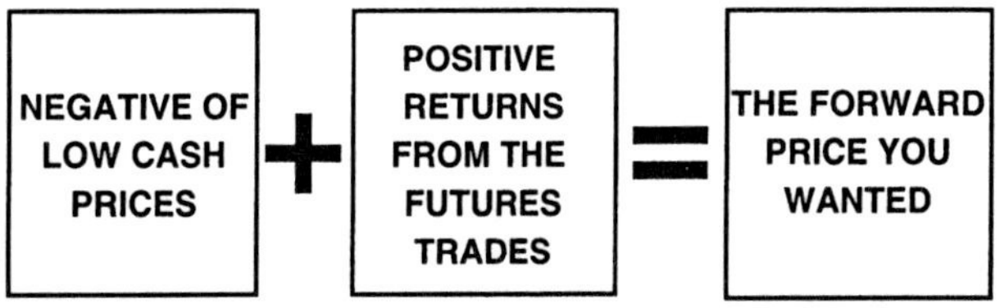

When prices fall, gain in futures offset low cash prices

Basis risk

The reader needs to keep in mind that once the forward price of $2.80 is established, that is exactly what the farmer will get no matter whether the overall price structure goes up or down so long as the difference between the two markets at harvest is the expected $.10 per bushel. It would be very interesting and useful for the reader, for example, to think through what would happen if, instead of the market going down, dry weather pushes corn prices up because yields are bad and the product is sold in the cash market at $3.10 per bushel instead of $2.10. But if the cash market is at $3.10, we would expect the futures to be around $3.20. Assume, again, that historical data suggests a basis of $-.10 per bushel. The producer would have to buy back the futures at around $3.20, and that constitutes a futures account loss of $.30 per bushel. Now, in getting net price, we combine $3.10 from the cash market and a negative -$.30 from the futures market. We have $3.10 - $.30 = $2.80. *The producer is guaranteed his $2.80 whether the market goes up or down so long as the basis performs as expected and moves to the -$.10 per bushel level.*

POSITIVES OF HIGH COST PRICES − **LOSSES FROM THE FUTURES TRADES** = **THE FORWARD PRICE YOU WANTED**

When prices rise, losses in futures offset unexpected gain in cash

We could go through more examples. A business firm that is using corn or some other agricultural commodity as a raw material, may be much more concerned about their costs going up than they are about the price level going down. This would be what we call a "long hedge." Instead of selling some distant futures contract, the user of the agricultural raw material would *buy* that distant futures contract. That way, if the cash market goes up as the manager of the processing firm might fear it would, then the futures market will go up with cash. That basis level or differential between cash and futures is going to be involved again. The long hedger would offset the rising cost in the cash market by selling the futures position at a price higher than the level at which they bought it. The net cost that was established when the framework was setup will be realized.

IMPORTANCE OF FUTURES

Going through an exhaustive set of examples is not the intent of this chapter. The important thing we need to get established is that all of this is made possible because there is trade in futures contracts. If there were no trade in futures contracts, there would be no place for the processor, the grain elevator, or the meat packer to offset their price risk, and they would not be willing to offer cash contracts calling for later delivery. If they offered them at all, they would offer only sharply lower prices to the producer because they would be worried that they would establish a price and then have that commitment in place while the overall price and the overall value of the inventory they are buying went down. All of this process of forward pricing is made possible because there is a futures complex. That is the primary need that must be established here. The direct use of the futures, as we have indicated above, establishes a specific price. The only thing that can keep that specific price from being realized exactly is if the basis at harvest, or whenever the operation is closed out is different than the one that had been expected. It would be very useful, for example, for the reader to go back to Table I and think about what would happen if, when the cash corn is sold at $2.10 on October 14, the best the producer could do would be to buy back the December futures at $2.25. Now, we have a net price at the bottom of the table of only $2.75, and we have seen what is sometimes called "basis risk" in the business. There is no guarantee that the basis in October of the next year will be at the same level that it tended to average across the past five years. This component of overall risk exposure or basis risk, while small relative to fluctuations in the cash market, is in fact still present.

If the producer is concerned about the basis risk and about know-how in terms of dealing directly in the futures market, then we have already discovered that he can go to the grain elevator or the meat packer and negotiate a cash contract for forward delivery. In effect, this means that the buying processor or packer takes the positions in the futures market for the farmer.

FORWARD PRICING IN FUTURES = **TRANSFER OF EXPOSURE TO PRICE RISK**

One reason futures trade exists is to allow transfer of price risk

There are some costs involved when the producer goes directly into the futures complex. Actually, those are the same costs incurred by the grain elevator or the buying meat packer who offer cash contracts. They are just passed on to the producer in the form of a lower cash contract price offer. The futures contracts are for a specific size. A corn contract, for example, traded on the CBOT is 5,000 bushels. The wheat contracts there are also 5,000, as are the soybean contracts. The slaughter cattle contracts traded on the CME are 40,000 lbs. or 400 cwt. There is a substantial value in each of the contracts, it is apparent, and trading those is a service for which the producer will have to pay a brokerage fee. It is not a great deal different than buying or selling a piece of real estate where a commission or brokerage fee is paid to a real estate broker. There are commodity brokers who can do this for the producer, and each buy and sell or each subsequent sell and then buy, which is called a "round-turn" in the business, carries with it a commission cost. Commissions on a 5,000-bushel corn contract can range anywhere from $30 to perhaps $80, which means something less than $.01 per bushel to perhaps $.015 per bushel depending on which brokerage firm is employed. Discount houses that offer few services will trade the commodities for you cheaper. Other more nearly "full service" brokerage firms will charge a bigger commission fee.

Commission fees to your commodity broker are one cost of trading futures

TRADE IS ON THE MARGIN

The only additional cost is a margin of roughly 5 percent of the face value of the contract has to be on deposit with the broker before you can trade futures. If the market moves against you, such as moving up toward the $3.20 per bushel discussed above, added margin money will be needed. If you have sold at $2.90 or taken a short position at $2.90, there would be losses incurred in the futures account as at the market starts to move up. The producer will have to send more money to the broker, something called the infamous "margin call." The producer really should charge interest on these margin funds as a cost of maintaining the futures positions. The total cost would be from $.01 to $.03 per bushel on grains and oilseeds and $.10 up to perhaps $.40 per hundredweight on slaughter livestock. For these and other agricultural commodities, the costs tend to be nominal in comparison to the face value. For example, a $.01-.02 cost on a wheat contract where the price is $3.00-4.00 per bushel or on a soybean contract where the price may be $5.00-7.00 per bushel is really of nominal importance and is not a significant obstacle to any producer using these markets.

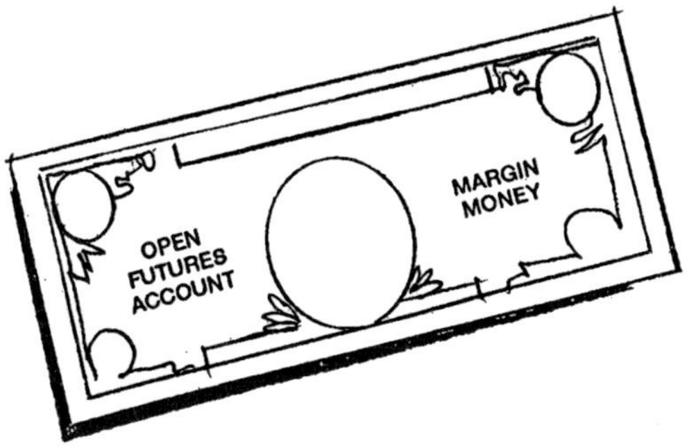

Margin money is needed to trade futures contracts

USE OF THE PUT OPTION

As suggested earlier in this chapter, options on agricultural commodities are relatively new. Most experts believe that they have added a useful, interesting, and exciting (at least to many producers) alternative to either the cash contracts or the use of positions directly in futures.

From the viewpoint of the individual producer, options provide a significantly different approach to forward pricing. Specifically, the option:

- **Provides for setting a price floor for the producer instead of establishing a specific price.**
- **Provides a clear indication up front as to what the cost of the program will be with no unexpected additional costs in the form of margin calls, etc.**
- **Leaves open the possible benefits of unexpectedly higher prices should they occur.**

Options give the producer a price floor and leave open the possibility of higher prices

Focusing on the put option, which is the option approach that parallels the short position directly in futures, the difference between the two can be shown by a formula that provides for a floor price:

FI P = SP + BASIS - PREM

where:

FI P = Floor price established by buying the put option.

SP = A "strike price," which is a specific level of the underlying futures price.

BASIS = The cash minus futures differential basis level as introduced earlier.

PREM = The premium or cost of the put option to the producer.

The exchanges offer options for various increments called strike prices; $.10 intervals in corn, for example, and $.25 intervals in soybeans or, in slaughter livestock, typically $1.00 per hundredweight intervals. In other words, the producer has to pick one of those incremental prices and cannot pick, for example, a $2.875 option on corn. He would either have to go to the $2.80 option or the $2.90 option.

Premium costs are the added costs of using options

The only other new addition in the floor price formula is the premium cost. This really amounts to paying someone else to accept the exposure to the futures market and the risk that comes with it. The treatment here is elementary, and we need not get into all the intricacies of what makes up the option premium. It is sufficient for us to recognize that the longer the time period involved, the higher the premium will be. This is called "time value" of the option in the business. What this means is that if you are buying an option at any particular strike price level on the December futures in April, it will cost you more if the price levels stay the same than would that same option in terms of price level in August. The longer the time period, the more risk exposure the person offering the option will have, and the bigger the premium will be. These premiums are determined in the competitive marketplace in or near the futures pits much like the futures prices are discovered.

The other major factor in terms of how large the premium will be is the volatility in the market. When the markets are highly volatile and risky, the person offering the option will want more pay for that service. The option premium tends to go up.

SHORT HEDGE FORWARD PRICE − **PUT OPTION PREMIUM** = **FLOOR PRICE USING THE OPTIONS**

True so long as the strike price selected equals the futures price in the short hedge

To demonstrate a put option using the developments in Table I, we could get:

FI P = SP + BASIS - PREM

= $2.90 - $.10 - $.20

= $2.60

This suggests that the floor price is made up of a strike price of $2.90 to correspond to the $2.90 we used in the short hedge example, less the $.10 basis (because basis is negative), less a $.20 premium cost. If we are looking from April out into October and the underlying December futures is, in fact, trading around $2.90, this is what is called an "at the money" option. It may cost as much as $.20 per bushel to get someone to offer that option to you.

FLOOR PRICE ≠ **SHORT HEDGE PRICE**

Depending on strike price level, floor prices using options will usually be less than the short hedge price

THE FUTURES AND PRICE VARIABILITY

The big difference is fairly easy to see. Let's assume that the market goes up substantially to, say, $4.00 per bushel because of a widespread drought and because production and supply has been reduced substantially. The producer who uses a straight hedge would still get the $2.80 price as shown in the table. The cost of this hedge would go up because interest on more margin monies would be involved. As the cash market goes up toward $4.00, the futures would go to something above $4.00. *Those losses in the futures don't come back to the producer.* He has a tremendous gain in the cash that is essentially offset by the losses in the futures. The windfall profit that would be there to the cash market speculator who had done no forward pricing would be denied to the hedger. Said another way, a hedge does not distinguish between positive and negative moves in price. It protects you against *all the price variability*, including the higher prices.

When prices rise, buying a put option is like buying insurance you did not need

Buying the $2.90 put option makes that scenario significantly different. If the futures market goes up to $4.10, for example, because of a drought, then the producer just allows that $2.90 option, which was the right to be short in the market at $2.90, to expire. He spent $.20 on price insurance that, as it turns out, he didn't need. He nets $4.00 plus the -$.10 basis (since he is selling the cash commodity in his local market), minus the $.20 option premium, and he will end up with something like $3.70 when he sells his corn in the cash market. The arithmetic is technically $4.00 + (-.10) - .20 = $3.70. Notice that is nearly $1.00 above the specific price that is made available to the producer who uses the straight hedge.

If the market goes down, the reader might be wondering how the option protects and ensures that the floor price will be realized. Let's go back to the table example where the cash price goes down to $2.10. That means, as we suggested, that the December futures should be around $2.20. *The right to be short at $2.90 when the futures are down to $2.20 should be approaching $.70 per bushel.* So what we have then is a $2.10 selling price, plus a $.70 total value of that option when it is sold at $.70 minus the initial option premium of $.20. So we have $2.10 plus $.70 equals $2.80, minus the $.20, and we are back to the $2.60 floor price. So long as the cash-futures basis does, in fact, go to the -$.10 as expected, that $2.90 put will pick up value. The option can be sold at its higher value, and the floor price of $2.60 will be assured when you combine the cash selling price, what you get from selling the option, and deduct the initial option premium.

PRICING STRATEGIES

CASH CONTRACTS	FUTURES HEDGES	OPTIONS ON FUTURES

We have the building blocks for pricing strategies

SUMMARY

There are, as implied, a number of ways to forward price. You can do it with a cash contract, you can do it directly in the futures, or you can use options. There are many possible demonstrations and illustrations that can be completed. We haven't really illustrated in this chapter, for example, exactly how it would work for a long hedge when the concern is about the cost of a raw material going up. Those possibilities and alternatives will be demonstrated throughout the book. The intent and need here was to stress to the reader the fact that without trade in futures, none of this would be possible.

The futures markets discover a price for a future time period, and in the process of doing that, they create a mechanism that allows the farmer to transfer exposure to price risk to someone else. The relatively new options bring an exciting addition to the farmer's kit of price risk management tools. Historically, farmers have been reluctant to use the futures market directly. They always fear the infamous margin call,

RIGHT TO BE SHORT AT $2.90 → **FUTURES PRICE GOES TO $2.20** → **GROSS PROFIT OF $.70**

A $2.90 "put" will approach a premium value of $.70 when futures go to $2.20

and they are always reluctant to peg a specific price level because there is always the possibility, albeit a small one, that the market will rally to substantially higher levels. The option eliminates concerns about a margin call, makes clear what the cost of the price insurance program is up front, and it leaves open the possibility of higher prices if they should occur. All this means that farmers who want to get involved in forward pricing have a market and an array of price risk management tools at their disposal to do something about their position as price takers.

In the later chapters, we will demonstrate all of these approaches and look at how to decide which approach to use. There are a few general factors that will always be present:

- **The cash and futures markets are discovering prices for the same commodity.**
- **The futures market also discovers prices for later time periods.**
- **Cash contracts for later delivery are always tied to the futures market and would not be available if there were no futures markets.**
- **Costs of any of the approaches to forward pricing will be typically nominal.**
- **The cash-futures basis is always involved, and the producer who prices in futures or uses options is exposed to basis risk.**

With this background, we have a base on which to build pricing strategies.

5

Basics in Managing Price Risk

INTRODUCTION

Discussions in Chapters 1 through 3 identify a very real need to do something about price risk. It appears that the individual producer as a price taker cannot avoid, on occasion, having to sell product below costs of production. In discussing seasonal patterns in agricultural commodities, price cycles across years, and the micro-macro trap that lurks for the individual producer, the inability to control price was always there. Obviously, it would be nice to take advantage of hard-earned awareness that a cyclical downturn in hog prices is likely to occur within the next few months. Having access to some way to establish a selling price for hogs that will be sold as much as a year in the future takes on a lot of importance. There are such possibilities, of course. Chapter 4 introduced the very basic framework, and this chapter builds on an extends that coverage.

Producers can't stop price form dropping below costs

WHEN PRICE IS SET

We want to emphasize that the one thing the producer can control is when price is established. By sharpening abilities to understand price patterns and by being prepared to forward price when the circumstances appear to be right, the individual producer can do something to improve his or her situation. By controlling when price is established, those huge losses that can come up on occasion might be eliminated or at least reduced. The notion of **forward pricing** of agricultural commodities is built on the recognition that the producer can control when price is established.

FUTURES AND SEASONAL PRICES

Before looking at the components of forward pricing, we have to stop and make sure we understand the real world. Being able to forward price is not going to turn the difficult, risky, and volatile agricultural markets into stable and non-risky markets. Even with the know-how to practice forward pricing, protection is never going to be perfect. In Chapter 2, there was considerable discussion about the substantial movement of slaughter hog prices within the year. The seasonal index suggested that hog prices tend to be highest in the summer months, especially July, and low in the fall months, especially October and November. If slaughter hog prices are at a very profitable $55 per hundredweight in July, this does *not* mean that even the astute producer will be able to forward price hogs to be sold in October at that same $55 level. *We have to understand that the marketplace that is discovering a price for future delivery has some awareness that there is a seasonal pattern in hog prices.* Thus, if you look at the prices being discovered for July delivery and for October delivery from a vantage point of, say, early June, the $55 price for July is likely to be accompanied by a $48 or $49 price for October. There is no way, then, to get the seasonal high hog prices year round.

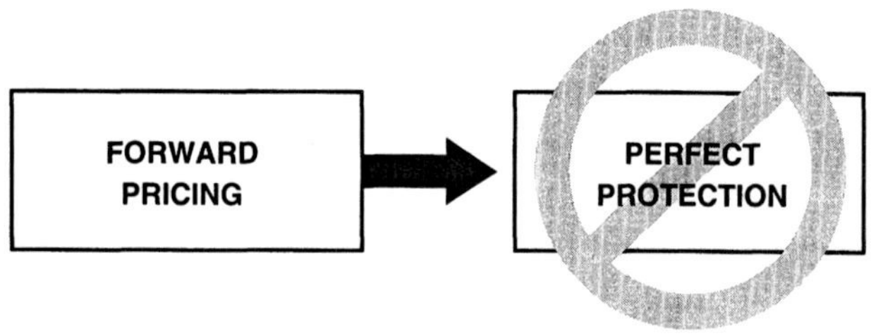

Forward pricing does not guarantee perfect protection

Not being able to get the peak prices does not mean that forward price will not work, however. During June and July, when the cash hog prices tend to climb as daily slaughter levels get reduced, the July futures will start trying to anticipate just how high the cash market will go this particular year. In late May, the cash hog market may still be $49 per hundredweight, but the discovered price for July may already be at $53 and climbing. If that price for July delivery climbs to $55, there is a high probability that the October price will climb as well. It may not climb to $55, but it may climb to potentially the best level you will see for the October delivery period all year. Thus, the producer may be in a position when cash hogs go to $54 or $55 in July, or even prior to that time period, to forward price October hogs at, say, $49 per hundredweight.

That $49 is not as good as $55, but it is profitable to an efficient producer. It will look very good if the seasonal increase in hog slaughter during October and November is especially strong in this particular year and prices are driven down toward $41-42, about break-even for many producers. *Forward pricing is not perfect. If it were that easy to get the highest price at all times, production agriculture would be flooded with people who understand how to do this.* Such a situation would be self-defeating. Supplies would be so high that price would be driven down and everybody would suffer. This latter point leads us also into another brief discussion of the micro-macro trap.

It happens in hogs; it happens in surges in planted acreage in corn; it happens in production of any agricultural commodity. When prices are relatively favorable, producers tend to get excited and overexpand. This is the micro or firm-level reaction that, when added together, moves the industry or macro-level supply curve and moves price. For a given level of demand in the international market for soybeans, for example, you can be sure that surging to 63 million planted acres in any particular year (because year before there was a drought and soybean prices went to $8 per bushel, a very profitable level) is going to guarantee that prices will be driven back down. If 60 million is normal, 63 million acres will drive prices sharply lower. This situation is not like being in May, anticipating seasonal strength in hog prices, and trying to forward price the hogs to be sold in October or November while prices are still good. Here, it is a matter of seeing that drought has reduced production and you are going into the harvest of soybeans in mid-October and prices are at $8 per bushel. *You can predict with almost certainty that acreage is going to expand the next year, supply will increase, and price will be driven back down.* Thus, it may be that October of year 1 may be the best time to forward price soybeans for the next year, year 2. *This pricing action would need to be considered six months before the year 2 crop is even planted!*

Higher prices bring increased supplies even if producers have to borrow to expand

TIME CAN WORK FOR THE PRODUCER

What all this does, of course, is makes time an element that can work to the benefit of the producer. If you know how to forward price and are comfortable with that process, then what you need to do is monitor what the market offers and then "pull the trigger" and establish a price when it reaches an acceptable level. Then, it is very important not to second guess yourself. On occasion, you will establish your price on one day only to watch the opportunities to price still higher emerge in the following weeks and months. It is an uncertain business. *You should never see as a mistake an action that was taken to establish a profitable price, get your exposure to price risk under control, and protect the economic viability of your program.*

COMPONENTS OF FORWARD PRICING

As noted in Chapter 4, there are always certain conditions that have to exist for the producer to be able to forward price. There has to be some organized market that is discovering a price for delivery in some future time period. The visible markets in the United States for agricultural commodities are the futures markets that were introduced in Chapter 4. We will come back and deal with those markets in detail later. But the important point is that there has to be some marketplace that is discovering price for later delivery if you are going to be able to forward price.

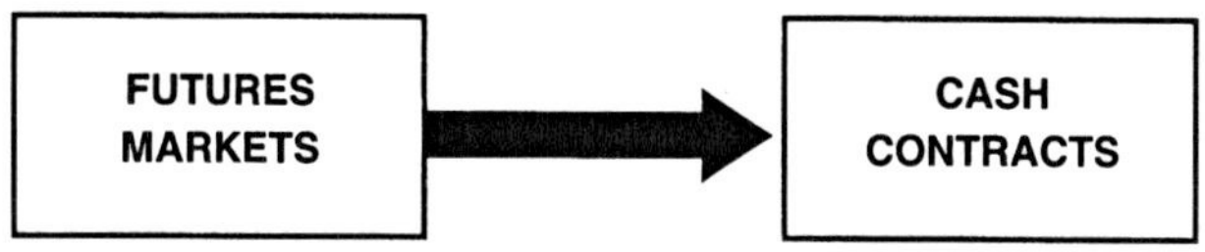

We can have cash contracts because we have futures markets

AVAILABILITY

This was introduced in Chapter 4, but it is very important that the reader understand that this need for a "market" is there even though he may be negotiating a cash contract with a local grain elevator. The manager of the grain elevator, who is willing to forward price wheat via a cash contract for delivery in the first two weeks in June, would not be able to extend a decent price offer if the elevator had to carry the risk that price would decline dramatically before the wheat is delivered. Obviously, no buyer would like to carry the risk if in early January they offer $3.50 per bushel for wheat to be delivered in the first two weeks of June, then actually have that wheat delivered in early June in a market with the price down to $2.50. There is no way you can be effective enough as a merchandiser of wheat to overcome the losses associated with having paid $3.50 for wheat that turns out to be worth, in early June, only $2.50. What happens, of course, is that the grain elevators are able to offer the $3.50 price because they, in turn, take a position in the futures market to protect themselves in the event that price goes down. We will find, then, that most cash contracts are just an indirect expression of the price being discovered in the futures complex. We will revisit this later when we talk about different ways of forward pricing and compare the alternatives.

Another component of forward pricing that has to be present is the ability to localize the prices being discovered for a future time period. In the example we discussed above with the wheat contract, the grain elevator has already localized the discovered price at the Chicago Board of Trade in the $3.50 offer for a cash contract. The price discovery process is often at one location. In the U.S., the price discovery process for grain, oilseeds, and livestock futures (as well as financial futures and many of the instruments employed in the world business arena) are located in Chicago. Other commodities are traded in New York City, and certain types of wheat are traded on exchanges in Kansas City and Minneapolis.

PRICE VARIES BY LOCATION

The point is not the city where the trade is occurring, but that the price that is being discovered for some future time period is almost never directly representative of your local market. Thus, there has to be an understanding of the need to localize the New York, Chicago, Kansas City, or Minneapolis quote and to be able to decipher what it means to you in your local market.

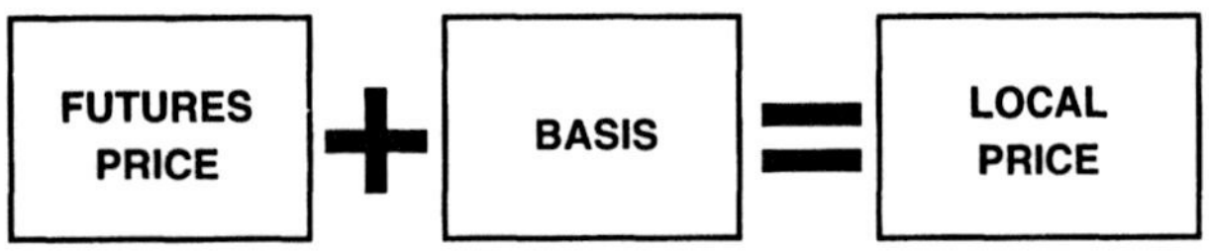

The basis adjustment converts the futures price to a local price

BASIS CONVERTS TO LOCAL MARKETS

This brings us to a very important component of any forward pricing program: **the cash-futures basis**. *The cash-futures basis is employed to localize a futures price that is discovered in, say, Chicago down to a particular location in the delta of Mississippi, in central Illinois, or in eastern Colorado.* This is the adjustment to southern Illinois that was introduced very briefly in Chapter 4. It doesn't matter where the location is; there is a way to pull that Chicago-based quote down to local conditions and understand what it means for a particular producer.

Basis is defined as cash minus futures. This means that for most commodities, for most periods of the year, this basis will be negative. There are some exceptions, but cash corn in southern Illinois is usually considered to be worth less than corn in Chicago. It is true of corn, cotton, soybeans, slaughter cattle, and slaughter hogs. Cash prices for these products are typically *below* futures, so the basis is typically negative.

Records on historical basis levels are important

That does not, of course, bother us in using the concept. Typically, the individual producer, the local extension agent, or the university specialist for different regions in your state will have a record of cash and futures prices back across the past five years. Those data are analyzed and converted into a table. The format is shown in Table I, and it is important that the reader understand what is involved.

Calendar Months	Futures Months (\$/bushel) March	May	July	Sept	Dec
January					
February					
March					
April					
May					
June					
July					
August					-.08 -.18 to -.06
September					-.10 -.20 to -.03
October					-.15 -.31 to -.07
November					-.11 -.22 to -.06
December					-.10 -.19 to -.01

Table I - Format for reporting the cash-futures basis for corn

Across the top of the table are the months for which corn futures are traded. Note that futures are not traded for each month. The futures months are March, May, July, September, and December. Down the left side of the table are the calendar months in any particular year. What the table becomes, then, is a matrix of basis levels for certain periods of the year *for your particular market*. This needs and deserves some additional explanation--pay close attention here!

DECEMBER FUTURES ARE IMPORTANT

In Table I, the focus of attention is on the basis for various calendar months against the December futures. If the entire table were filled, it looks imposing in terms of its apparent complexity. It is useful here just to simplify and focus on a certain part of the table that a producer who is thinking about forward pricing corn will need to use.

In this particular market area where our producer is located, let's assume that the bulk of the corn harvest is typically in late September and early October. Notice that across past years (let's assume it is the most recent five years), cash corn prices in his market area have averaged $.15 per bushel below the December futures during early October, with a range of from -$.31 to -$.07. These are the very important data that the producer needs to localize the futures quote for his particular market area.

A specific basis level has to be selected

Let's assume that the producer decides to use the average of -$.15 to adjust the Chicago Board of Trade quotes. We may have occasion to come back later and think about why he might choose some level other than the average, but to keep it simple, let's assume that the -$.15 is used. *Specifically, this means that across the past five years, selling prices for corn in the producer's local cash market have averaged, during October, $.15 per bushel below the trading level of the December futures during October.* These data could have been generated several different ways. One simple way to think about it is to average the cash quotes for the entire month of October across each of the past five Octobers, then average the closing price for the *December* corn futures for each day across the past five Octobers. Subtract the average futures from the average cash, and we get the -$.15. In practice, some producers will use a mid-week cash quote and a mid-week close in the futures just to simplify the calculation. Each year, new data, both cash and futures prices, could be added, and the oldest year could then be dropped. That way, a five-year average for the basis is maintained in the table. Obviously, most producers would want to do this using a computerized spreadsheet program or get updated tables from a local extension agent or the land-grant university specialist.

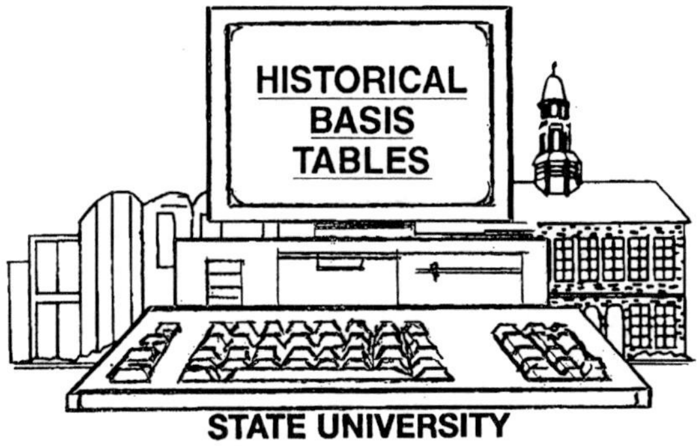

Your land grant university specialist can help with basis data

CLOSING BASIS ESTIMATE IS USED ALL YEAR

Once the table has been updated, the producer now has what he needs to localize the price for December corn futures all through the year. We will show in repeated examples that you always go to a futures month *after* the time you expect to have your cash product ready to sell. The December corn futures is the new-crop corn futures contract for most corn producers in the United States. Once the -$.15 is calculated, the producer can monitor the closing price for the December corn futures in the newspaper or via electronic services that are widely available at very low costs. Each day, mentally subtract $.15 from the futures quote and you have an indication of what the futures market is offering you as a forward price. *It is not important that procedure on how to forward price be understood in great detail at this point in time.* The coverage in Chapter 4 and the discussion here is building toward that understanding. What *is* important is that the reader understand the importance of basis information and how basis information, once generated, can be used to localize the price being discovered at the Chicago Board of Trade.

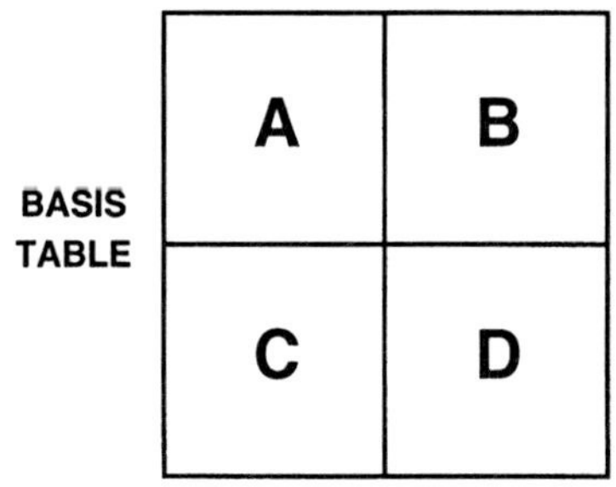

Different parts of the basis table are used for different decisions

Whether it is cattle, hogs, a storage program for soybeans, a growing program for cotton, etc., the producer will need a complete basis table following the format established in Table I. Different parts of the table will be employed depending on what decision is being considered. The discussion to this point has focused on a producer who is getting ready to plant a corn crop and is thinking about how he might be able to forward price it directly in the futures market. Later, we will find that another important decision has to be made in October. The producer has to decide whether or not this a year in which using any available on-farm bins to store the corn looks like it would be a profitable alternative.

We will come back to this storage decision later, but what we will find is that a different part of Table I will have to be used. The producer will then be most interested in what the cash-futures basis is in October using the May corn futures. He will also be interested in what that basis is likely to do as we approach May, and we will find that basis patterns are the key to an intelligent and, over time, profitable set of storage decisions. Detail will come later. To repeat the important point here, the producer must have access to the type of basis information shown in Table I.

TYPES OF FORWARD PRICING

As noted briefly in Chapter 4, there are three broad approaches to forward pricing:

- **Cash contract for later delivery**
- **Positions directly in the futures**
- **Options on futures contracts**

A specific approach to forward pricing has to be selected

The most familiar to producers is a cash contract. A contract with a specified price is negotiated with the local grain elevator for corn, or with a soybean processor in the neighboring county who might be interested in buying your soybeans, or with a nearby swine slaughtering operation who is interested in contracting with you to buy your slaughter hogs. *This is the most widely used type of forward pricing for the major agricultural commodities in the U.S.* As suggested earlier, *it constitutes an indirect use of the underlying futures market.* There has to be some mechanism, we said, to discover price for a future time period before the buyer is willing to offer a cash contract for future delivery.

CASH CONTRACT

The cash contract will specify, (1) a particular price as negotiated between buyer and seller, (2) a time period in which delivery is possible, (3) any quality characteristics that the buyer might insist on and the seller will agree to, and (4) who has the right to control the specific delivery date within a specified time period. Normally, the buyer will reserve this latter right. It sometimes gets to be controversial. Packers may call and ask for a set of slaughter hogs to be delivered early in the 10-day time period specified in the contract, and the producer might prefer to be able to hold those hogs and delivery them later to enjoy the benefits of the added weight. But without getting into the subtleties of the program at this point, the important thing is to understand what a cash contract is and is not and how it can be used.

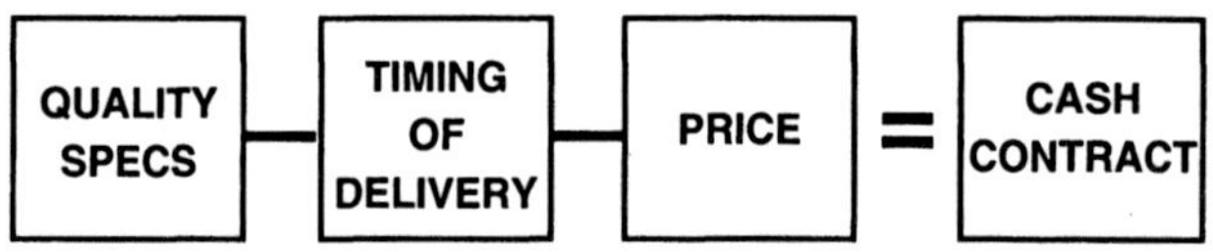

The elements of a cash contract

The cash contract, then, has several specific characteristics:

- **It is a written contract signed and dated by both buyer and seller.**
- **It will typically indicate a specific cash price to be received when the product is delivered.**
- **There will usually be specific quality characteristics that have to be met. In grain contracts, for example, this will typically include price premiums and price discounts for foreign matter, for moisture level in the grain, etc.**
- **It will typically include a "time window" during which the product can be delivered, showing a beginning date and ending date.**
- **The contract will usually specify who has authority to designate the particular delivery date within the contractual delivery period. Increasingly this right rests with the buyer.**

The cash contract is much like a contract that involves the sale or purchase of a house, an automobile, real estate, land, investment property, etc. It specifies the important legal dimensions of the contract, and establishes who has responsibility to perform certain acts, as well as how any adjustments are allowed relative to the contract provisions.

Making a decision on forward pricing is a must

The producer's interest

Of particular interest to us here is the price that is established in the cash contract. Consistent with the discussion above, this price is typically determined by the buyer after "localizing" the futures quotes. For example, let's assume we are dealing with a grain elevator who is interested in a cash contract with a producer for delivery of corn between September 25 and October 10. The date is June 10, and both buyer and seller are interested in getting price established. The producer as a seller has the obvious concern. Price might be lower at harvest than is currently being offered, and price might fall below cost of production. Remember, the producer is a price taker!

The buyer's interest

The buyer's motivation is different. Typically, the elevator manager is interested in making sure that he has an adequate supply of corn of a certain quality to meet processing, feeding, or exporting needs. Thus, there is a motivation to try to "tie down" some of that needed supply for that late September or early October time period prior to the actual harvest. A grain elevator will offer on each day two types of price offers. One is for corn to be delivered that day, in this instance, June 10. The second would be what is called a **"new-crop bid"** or a bid for corn to be delivered at some designated part of the upcoming harvest period.

Let's assume that the December corn futures are trading at $3.00, and the grain elevator is offering the producer a cash contract at a price of $2.75 for late September, early October delivery. That $.25 differential between the December futures price and the cash price offered by the grain elevator reflects the elevator's understanding of the relationship between this particular market area and the discovered price in Chicago. In other words, it reflects the basis level being used on this particular day by the grain elevator to establish a cash price. That *basis* relative to December futures and the elevator's offer can and will change over time. If it is -$.25, as implied here on June 10, it might be something different on July 15. On July 15, the grain elevator manager knows what the weather situation has been at least part-way through the growing season, knows what the 30-day weather forecast is, and is anticipating a substantial increase in the crop and lower prices in the fall. On July 15, the December corn futures may be down to $2.75, reflecting the expectation of an increased crop. But there is no guarantee that the bid for cash contracts being offered by the grain elevator will be at $2.50. The manager is now more comfortable that he will have all the corn he will need in September and October. He may expand the basis that is employed to, say, -$.30, which means the cash contract bid might be $2.45, not $2.50, when the futures are at $2.75. That change in basis is a direct expression of the changed outlook by the grain elevator.

The grain elevation is looking for a margin

Futures offer an alternative

As we proceed to look at other types of forward pricing it is important that the reader keep in mind that one reason to be interested in the futures market as a place to forward price is to keep the bids by the grain elevator competitive. *There is always an alternative to the cash contract: and that is the futures market itself.* The grain elevators, after adjusting that December futures quote down to a cash offer, are going to use the December futures to protect themselves against declining prices if the cash contract is signed. What we will find is the producer can go directly to that December futures contract, and by entering the futures market directly instead of indirectly, forward price the corn himself. There is no reason, then, to take what appears to be an unacceptably wide or weak basis adjustment in the cash offer. The producer will have an alternative and can, in fact, compare the cash contract offer with what is being offered in the futures directly. As we proceed to sketch out the basic elements of forward pricing in the futures, this point needs to be kept in mind.

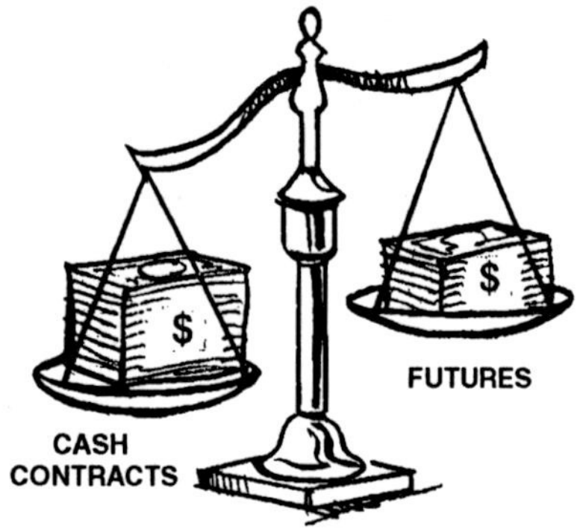

Compare cash contract and futures pricing

PRICING IN FUTURES

The second type of forward pricing *involves direct use of the futures market.* Keep in mind that the intent here is to briefly discuss the types of forward pricing and not to present details on "how to." We will come back later, especially in the next chapter, to talk about strategies and lay out detailed procedures for crops and for livestock. Here, we want to make sure the reader understands the alternative types of forward pricing available to them.

Forward pricing in the futures directly involves eliminating the middleman (the grain elevator, the hog packer, or whomever was issuing the cash contract) and going directly to the futures. Remember, when a cash contract is used, the buyer takes a position in futures. When using the futures directly, the producer takes that position in the futures.

The notion of a forward price, as it is being offered by the futures market, involves a futures trading level and the basis concept, which has already been introduced. It came into the discussion in Chapter 4 and earlier in this chapter. Specifically, to review, the forward price being offered by the futures market can be defined as follows:

Forward Price = Futures + Basis

or in terms of short-hand terminology used in Chapter 4,

FP = FUT + BASIS.

Note that the expression is futures price *plus* basis. This means that the futures quote of interest has to be adjusted by basis to get it to local conditions. By adding basis, the algebra will be right. Since basis is normally negative, when you *add* a negative number, you actually subtract the basis magnitude from the futures. But by expressing FP as futures plus basis, this also handles the more unusual circumstance where basis is positive. In that instance, you simply add the positive basis to the quote from the futures market to get the forward price being offered by the market.

Illustration

To illustrate the basics, let's assume that the December corn futures are trading at $3.00 on June 10. This is the same assumption that was used earlier when discussing how the grain elevator generates a cash contract. A producer goes to the basis tables and finds that for October the basis in this particular market area (the same one the elevator is located in!) has across recent years, averaged -$.22. The forward price the producer is being offered is, then, shown as follows:

FP = FUT + BASIS

FP = $3.00 - .22

FP = $2.78

Remember, the grain elevator was offering $2.75 for early October delivery. The producer finds that, according to his basis information, -$.22 is a reasonable basis adjustment. The price being offered directly in the futures is $2.78. The reader may well start to have a bit of concern creeping in here. It is fairly obvious where the price was coming from if you had taken the $2.75 cash price from the elevator. When you deliver the corn, the grain elevator is going to pay you $2.75 for each bushel. But if you forward price it in this somewhat more nebulous-appearing marketplace called the futures, *where is the $2.75 coming from?*

Futures and cash prices move together over time

Again, without getting too involved in the procedure, it is useful to keep in mind that the two markets, the cash market and the futures market, tend to move together over time. *Specifically, if the cash and December futures have tended to be $.22 apart in early October in past years, there is reason to expect them to be about $.22 apart again in this particular year.* We will find there is no guarantee of that, but the basis does tend to move to the same general level that it has shown historically. There is some variability (keep in mind the range suggested in Table I), but the general tendency is there. If we assume for the moment that the basis will, in fact, move toward -$.22, how does that help us?

What this means is that if the cash market goes down and is $2.30 a bushel in October, *the futures market will tend to go to a position about $.22 above the cash market, i.e., $2.52. This tendency for the two markets to move together over time as you approach the futures month for which trade is being considered becomes the essence of forward pricing in futures, or what is often called hedging.* If the cash market does, in fact, go down to $2.30, the producer has been hurt. He has passed up a cash contract bid of $2.75 and is now facing a $2.30 price. *This is what happens if you are a cash market speculator and do not take advantage of forward pricing opportunities!* Keep in mind, all of the discussion in previous chapters indicates that if in early October the price is $2.30, there is nothing the producer can do about it. He is a *price taker* on the day he hauls the corn to town, and *he cannot negotiate a higher price with the elevator just because he feels he needs or deserves it.*

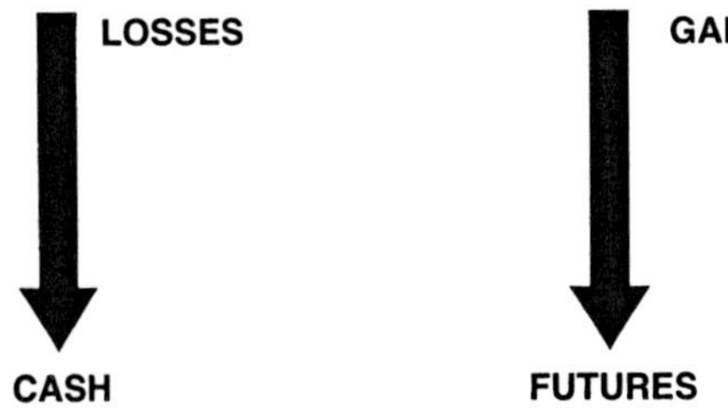

In declining markets, gains from short hedges in futures offset cash losses

The realized price

If the cash price is down to $2.30, then futures are likely to be down to $2.52. Again, without getting involved in detail, what we have is a futures position that was sold at $3.00 and can be bought back at $2.52. Selling high and buying back low makes money just like buying low and selling high, the old adage we have heard many times. The producer will be able to take essentially $.48 ($3.00 minus $2.52) less some nominal commission costs (remember the brief coverage in Chapter 4) out of the futures market. He adds that $.48 back to the $2.30 quoted in the cash market, and he has his $2.78 "forward price." This is a **realized price**, which is always made up of the best price you can get in the cash market plus the net you can get from your futures transactions. The basic notion is that the producer is protected because if cash goes down, futures will tend to go down too. Let's leave this point for later discussion and discuss a possible modification of the direct use of futures price quotes.

Formula price contract

One modification that is becoming increasingly popular but still involves use of the futures market directly is what is often called a basis contract or sometimes called a *formula price contract*. This means the buyer, such as the grain elevator or the hog packer, wants to secure a supply of product for a particular period in the future. But either he or the seller may not be interested in establishing price on this particular date. The seller may not be interested because he thinks better price opportunities might come along later. Still, the buyer is pressing him to commit the product so that he, in fact, will know that he will have the corn in early October or will know he will have the hogs at some later date. Under these circumstances, a basis contract or a formula price contract is often negotiated.

A formula price contract ties to a visible price series

What is involved is an agreement that prior to October 1, for example, the producer can pick any day he wants to establish price. He calls the grain elevator and indicates that he wants to establish price on this particular day, and the grain elevator manager then takes the necessary position in futures to protect the commitment that he has made. An agreement is made in the contract to a basis level or to some discount (or occasionally premium) to a futures contract. The producer as seller has the flexibility to decide when he wants to establish price. Obviously, what he would like to do is see that price go higher and then call and establish it at the higher level!

When to establish a price

One might argue that this variation of the futures market requires a bit more sophistication than does a direct selling of the futures which pegs a particular price, and it would be hard to disagree with that reasoning. The producer must remember that he has committed the corn or the hogs to the buyer, but he does not have any price protection when only a basis or a formula price contract has been signed. This means you have to be diligent, have to be observant, and have to watch what the market is doing. *Then, when you think the best possible price that you are likely to get is there, call and establish that price.*

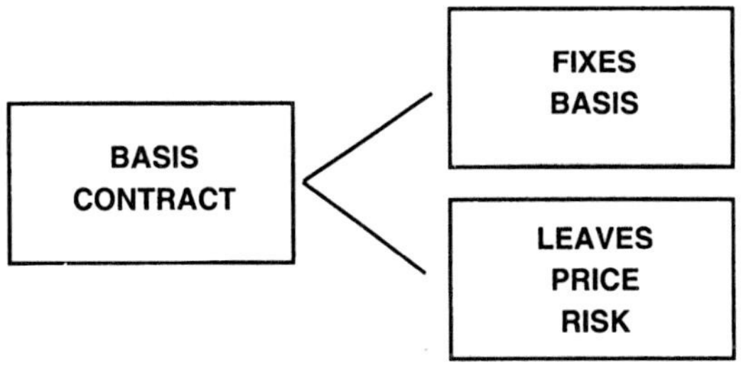

Basis contracts eliminate basis risk but not price risk

We can probably expect to see this type of contractual arrangement increase in the future. The buyer gets what he needs, a commitment to a supply of the product, and sellers can often get the flexibility that they like and make sure they have a place to sell the product at the later time period. The seller is obviously going to be disappointed if he passes up an opportunity for a $2.75 cash contract in corn, or a $2.78 forward price directly in the futures, in hopes for better prices later, and they don't occur. There is then going to be some lower final price established. If that price is never really set, you might only get the equivalent of that $2.30 per bushel price that the cash market speculator was forced to accept. This approach of leaving open the "when" of pricing takes discipline on the part of the producer.

A specific price

The basis contract will be demonstrated in the next chapter for both a growing program, whether it be for crops or livestock, and for a storage program in the grains and oilseeds. The contracts are relatively new (especially in livestock) and are still unknown to some buyers and sellers. But they offer some advantages in terms of flexibility that are appealing to many--both buyers and sellers. As suggested, we are likely to see their use increase.

One of the downsides of a cash contract or a forward price directly in the futures market is that it pegs a specific price level. That is apparent when you think about a cash contract, and we will increasingly come to understand that when you forward price directly in futures by selling the futures, you have established a particular price level. In Chapter 4, the illustration showed a specific price, subject to how closely the basis moves to your expected level. Many producers don't like this feature. They are especially concerned about establishing a price at $2.75 for corn and then watching the market, which, because of some unexpected weather disruptions during the summer months, may offer pricing opportunities for early October as high as, say, $3.30 per bushel. Somehow, they feel they have made a mistake by establishing a cash contract at $2.75.

The producers who established a price of about $2.78 by selling the futures directly may feel they are in even worse shape. When futures prices go up, not only are they restricted to that approximate $2.78 level, they have to send additional money to the brokerage firm that has taken the position for them to keep their protection in place. This is the infamous "margin call" mentioned in Chapter 4 that we will discuss later in some detail. They feel they have made a mistake initially by establishing price too low, and then, as the market goes up, have to pay somehow for the privilege of having made that mistake!

OPTIONS ON FUTURES

All this means that producers are increasingly finding favor with the third broad approach to forward pricing, options on futures. Not available for various reasons in the United States for many decades, options on commodity futures were launched again in the early 1980's on a trial basis as a way to help get the government out of the price support business. By the mid-1980's, options were being widely traded on virtually any agricultural commodity. *The options are significantly different than either the cash contract or the position in futures as a way of forward pricing.*

Types of options

Two types of options are traded. The **put option** refers to the right to sell the underlying futures at a particular futures price level called a strike price. The **call option** refers to the right to buy an underlying futures at a specific price level, a specific strike price.

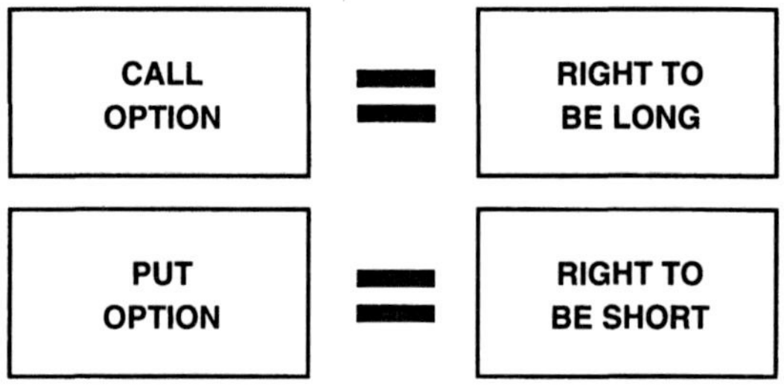

Options give "rights" to futures positions

The reader is reminded to proceed slowly here and not worry a great deal about this introduction of new terminology. There will be a number of demonstrations in later chapters and related discussion about when the options should be used versus the alternatives. In fact, we will start that discussion in very simple terms at the end of this chapter. The intent here is to make sure that there is understanding of how this third approach to forward pricing differs from the cash contract and from the position directly in futures.

To demonstrate, let's keep things simple and deal with the corn situation that has already been discussed. We have a producer who is looking at a cash contract opportunity on June 10 of $2.75. We found that after he adjusts the December corn futures for the normal cash-futures basis for early October, the futures market appears to be offering a $2.78 forward price. There is a caveat associated with that latter price in that it is predicated on the assumption that the -$.22 basis will occur again. But we have argued that those basis levels come to historical levels with a useful degree of accuracy. We have also introduced the idea that any producer who establishes a price at either $2.75 or $2.78 is probably going to have some concerns and start feeling like he made a mistake if the cash market does, in fact, go to $3.30. This is where the unique feature of the option comes in.

Price floor

The producer can use a put option to establish a price floor under the market. This allows the producer to get protection against the possibility that prices will decline significantly and leave open the advantages of higher prices if they unexpectedly should occur. This is going to need some additional explanation, so let's start that process by defining a floor price as follows:

Floor price = strike price + basis - premium

or in short-hand terminology as was first shown in Chapter 4:

Fl P = SP + BASIS - PREM

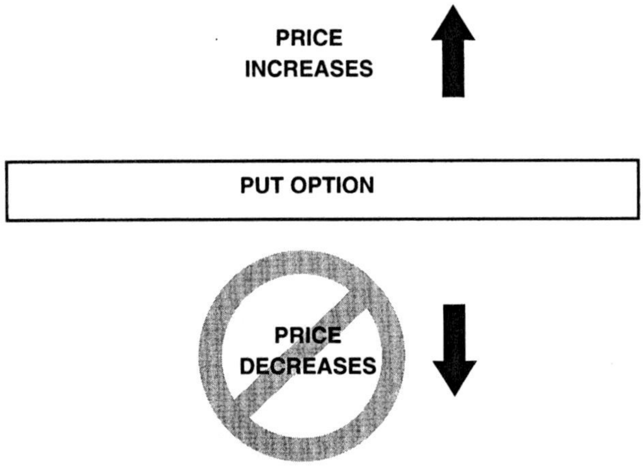

Put options only block lower prices

Illustration

To clarify how this works, let's assume that we know that the CBOT offers strike prices for corn futures in $.10 intervals. That is, you can buy a $2.80 or $2.90 put, or $3.00 put, or still higher $.10 intervals, if you prefer, when the futures are trading around $3.00. Later, we will talk about which strike price might be the best buy, but for the moment, let's assume that the $3.00 strike price is selected. For the right to be able to sell the underlying December futures at $3.00, let's assume that on June 10 we have to pay $.08 per bushel as a premium. We suggested in Chapter 4 that the premium cost of that "price insurance" will go up in very volatile markets, and it will go up when there is a longer time period between the date you are thinking about doing this and the maturity date on the option. There will be more on this later. For now, however, let's assume we can buy that $3.00 put for $.08.

Now the floor price becomes the following:

Fl P = SP + BASIS - PREM

or

Fl P = $3.00 - .22 - .08

Fl P = $2.70

Note that when we pay $.08 a bushel for the right to sell at $3.00, we establish a floor price, which turns out to be $.08 below the $2.78 direct forward pricing opportunity. The producer's initial reaction might be "why should I do this? Why don't I just take the $2.78?" But that initial reaction is overlooking the huge difference between the two approaches. Let's explain in more detail.

What we will find is that buying the $3.00 put option will offer protection in the event the price does fall down to levels such as the $2.30 in the cash market we talked about earlier. If a floor price of $2.70 is established using a $.22 basis allowance, then if that basis turns out to be $.22 in October, the $2.70 floor price will be realized. *What happens, as we will show later, is that when the cash price goes down, that $3.00 put option which was bought for $.08 will go up in value. You sell that option at its increased value and generate the money you need to take the $2.30 cash price back up to the $2.70 floor price.* You will get the $2.70 floor price exactly so long as the cash futures basis is, in fact, -$.22 when you sell your cash product and sell your option. But that is still a bit too complex and complicated for us at this point in time. What we want to emphasize here is what happens if the market goes up.

Put options let producers benefit from higher prices

Leaves upside open

Remember, the producer was upset when the cash market went to $3.30. A $3.30 cash price means the futures price is going to be near $3.52, given the -$.22 basis. If the cash market goes to $3.30, what does the producer with the cash contract get? He gets $2.75. What does the producer with the $2.78 forward price in futures get? He gets around $2.78, depending on whether basis is exactly -$.22 or not at the end of the program. What does the producer who bought the $3.00 put get? *He gets around $3.22 per bushel, and this is the very important difference to many producers.*

Worthless options expire

When the cash market goes to $3.30 and the futures go up to around $3.52 (remember, the -$.22 basis), the right to sell futures at $3.00 is worthless. What has happened is the producer bought price insurance that, as it turns out, he didn't need. He just allows this option to expire worthless and ignores the situation much as he ignores the premium he has paid for fire insurance on his home. *Any time the market goes strongly higher in such circumstances, the producer will net the cash price at which he can sell the corn minus the $.08 premium cost on price insurance that he does not use.* This offers the producer a huge advantage in those market periods when prices go unexpectedly higher. Instead of bemoaning the fact that he has priced his corn at $2.75 or $2.78 when it is, in fact, up to $3.30, the producer is elated over his wisdom! He recognizes that he had downside protection just in case, but he is getting the benefits of higher prices as well.

The right to be short at $3.00 is worthless when futures increase to $3.52

Minimum price contract

There is an alternative to the producer going directly into the market and buying the put option. This has cropped up in very recent years in the grain markets and is especially prevalent for corn and wheat. Elevator operators will now offer a minimum price contract. The same approach is starting to show up in the livestock commodities as well. What happens is that the grain elevator understands that they can buy that $3.00 put that we discussed above. They would pay the same $.08 for it that the producer would pay directly. They need to deduct the basis just as the producer would have done, and they then offer a minimum price contract, perhaps around the $2.70 that the producer could get directly, but perhaps down toward about $2.65 to offer a bit of additional margin for themselves.

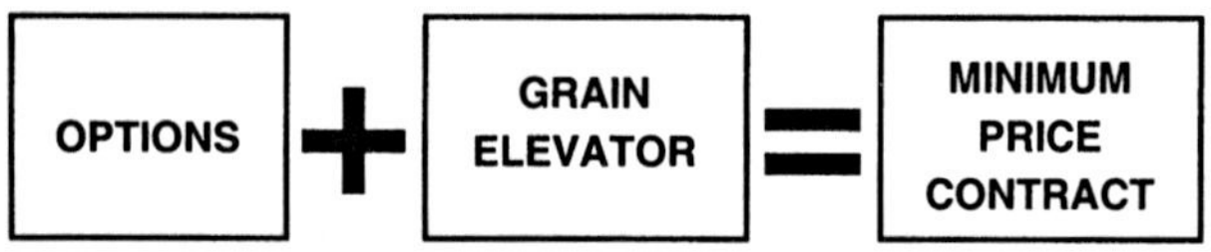

Trade in options has allowed buyers to offer minimum price contracts

The producer might very well be interested in taking the $2.65 minimum price contract offered by the elevator. Keep in mind that you don't have to pay the premium cost that way, that the elevator is incurring all of the cost of the premium on the put option, and that the elevator is incurring any risk that is associated with the possibility that the basis will not end up being the expected -$.22. (We will find later that if instead of -$.22 that basis should unexpectedly widen to -$.32, the $2.70 floor price that the producer thought he had established for himself drops down to $2.60.) By using the minimum price contract, the elevator faces that risk, and the producer doesn't have to worry with the detail. That is, in fact, one of the reasons why you will probably see the minimum price contract that the elevator operator offers a bit below the $2.70 price floor that the producer could get directly by buying the put option. Here, if price in the cash market was up to $3.30, the producer will get about $3.17--some $.05 below the $3.22 if he bought the put option directly.

Again, without getting too much involved in procedure, it is important to recognize that the option or the minimum price contract which is based on the option, does offer the producer a significant alternative. Instead of establishing a specific price in a cash contract or in a position directly in the futures, the producer has the opportunity to establish a floor price. This means there is protection if price does go down, and that producer still has open the possibility of benefiting from unexpectedly higher prices. It is apparent why this type of contract might be of particular interest to many producers.

WHICH APPROACH TO USE

In very basic terms, let's talk about the determinants of which approach a producer should use. As we develop case examples in later chapters, we will continue to come back to this point, but it is important that this general discussion of forward pricing deal with the important question of which approach.

Deciding which approach to use in pricing depends on economic and personal factors

There could be a long list of factors that will determine which approach a producer takes, but it would appear the most important ones would be as follows:

- **The financial position of the farm**
- **The risk aversion of the manager of this farm**
- **Where the price is at the time forward pricing is being considered, and the expected price range for the year**
- **The direction in which the price appears to be trending at the time price protection or forward pricing is being considered**

FINANCIAL POSITION

The financial position of the farm will be very important. Obviously, a position directly in the futures opens up the possibility that the producer will have to send additional margin monies to the brokerage firm. Later we will lay this procedure out and go through a case example that shows exactly how that margin process works. But it is sufficient here to recognize that if you sell the corn futures from the above discussion at $3.00 per bushel and then the futures go up, you will have to send money to the broker to cover the losses that you would incur if you have to buy these futures back at higher prices.

It is also important to recognize that the financial position of the farm has something to say about the farm's capacity to absorb risk. There may be instances in which no forward pricing is employed because the farm manager decides that he is just going to be a cash market speculator because the forward pricing opportunities don't look very attractive. That alternative may be there for a farm that is pretty well fixed in terms of financing. It may not be a realistic alternative for another that is heavily leveraged, has borrowed a lot money, and would be forced out of business if corn goes down sharply or if selling price for slaughter hogs plunges suddenly. In this latter case, it may make more sense to manage your financial commitment by going to the options. You know what it costs up front to get price protection. You can drop your strike price down, we will find, and reduce your cost of price protection. In the case of the minimum price contract offered by the grain elevator or the meat packer, they are paying the option premium for you up front and reflecting it in the lower floor price that they are establishing. Obviously, the financial position of the farm will have a bearing on which approach is considered to be appropriate in any particular set of circumstances.

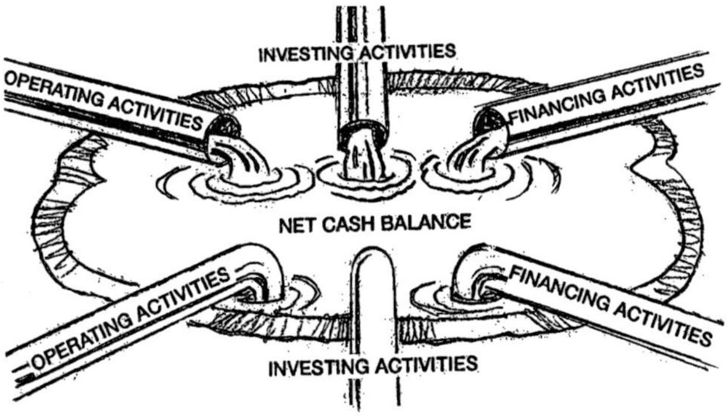

Your financial position affects your ability to tolerate risk

COMFORT ZONE IN RISK

There is no question that the degree to which the farm manager has aversion to risk is very important. Many managers thrive on the ups and downs of agricultural markets. Some farmers will tell you they would not be in business if they didn't have a "gambling" impulse at heart. But in a more business-like context, managers still vary in terms of how comfortable they can feel with exposure to price risk (or any type of risk) and whether or not they ought to seek protection.

A manager who is highly risk averse, for example, might prefer to establish a floor price early in the year by buying a put option and not have to worry about that decision for the rest of the year. He knows that he has established a price that is high enough to at least cover his out-of-pocket or variable cost of production, and he knows he has a price established that will enable him to pay off the loan at the bank come harvest time. He can then be the best producer he can be, get the yields as high as he can, get the gains as favorable as he can on the feeding floor for hogs or cattle he might be finishing, and not worry about the marketing side. Some other managers might take just the opposite approach: they might prefer to learn about the markets, develop expertise, and to argue that they are good enough analysts to be able to pick better prices later in the year. Thus, early in the year they may be strictly in a cash market speculative position and carrying all the risk exposure with their corn or hog program, but they intend to get some forward prices established later in the year. Upon reflection, it is apparent that the manager's attitude toward risk is going to have an influence on which way he or she goes in terms of forward pricing.

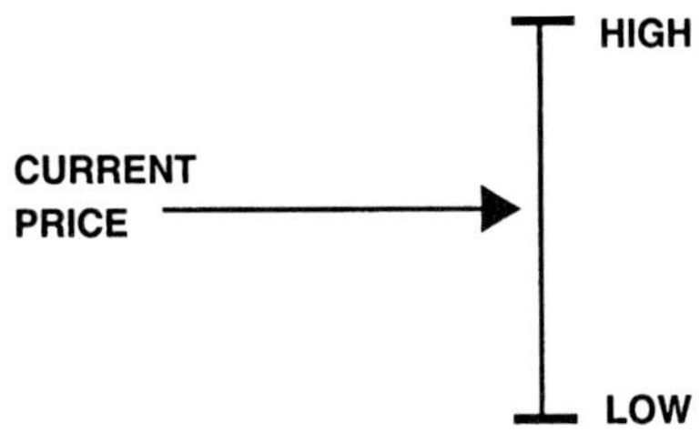

Where current price is in the range for the year will be important

PRICE LEVEL IN THE RANGE

Where the forward price being offered is in the price range is very important. For example, if a $3.00 December futures quote is near the highest price you have seen all year for that particular futures, it might make more sense to establish a cash contract or to go directly into the futures than to buy a $3.00 put. Buying a $3.00 put means you are paying a premium to allow upside flexibility. *If the price is already at or near the top end of the price range for the year, it has to establish a new high and move out of that price range to the upside before buying the put option will do you any good.* Conversely, if the price being offered is near the bottom end of the price range for the year and yet you need protection for a fragile financial situation or because the bank is insisting on it, then you might want to use a put option. Buying a put option when the price is near the bottom end of the expected price range for the year makes sense. You are buying upside flexibility that may well, in fact, be relevant in this particular year.

PROBABLE PRICE DIRECTION

The direction in which price is moving has comparable meaning to the decision on which way you go. If the price is trending up, if it appears that it will continue to trend up, and yet you feel you need downside protection, then buying a put option may be the more reasonable thing to do. If you really want to minimize your initial cash outlay, you might want to go to a minimum price contract with the local grain elevator. If, on the other hand, the price is trending down and apparently will continue down, it might make sense to establish a cash contract or go directly into the futures and sell the futures. If the trend does, in fact, continue down in the futures, you end up with a price that will be higher than the floor price established by using options. If you sold the futures and the price continues down, then there is no margin call. In fact, there will be profits accumulating in your account that you could pull out and bring back to the bank and use for other purposes. We will cover all this in more detail later when we go through the example on margins. But if price is, in fact, trending down and expected to continue trending down because it looks like you are going to have a huge crop or a huge supply of pork, then using a cash contract or going directly into futures might be preferable to options.

FORWARD PRICING = **MARKETING MANAGEMENT**

Forward pricing is the base on which a marketing plan is built

SUMMARY

Much of what this book is all about is forward pricing. In this chapter, we talked more about the three very basic approaches to forward pricing. A ***cash contract*** is the tried and true approach that has been around for a long time. Not many producers have always fully understood that you can have a cash contract only when there is some mechanism that is discovering prices for future time periods. In other words, the cash contract is always based on a futures price quote. There is nothing wrong with using a cash contract if this is the approach that the producer feels good about and is most comfortable with.

The **direct position in the futures** offers the producer an alternative. If the quote being offered in the cash contract doesn't look high enough or doesn't look competitive, then the producer can always go directly to the futures. Derivatives of this include the basis contract or the formula price contract that allows you to secure a market as a producer for your product and still have a time window of opportunity in which you are establishing price. A favorable basis has, hopefully, already been established. Using the futures directly or using a basis contract does, in fact, establish a certain price. However, once you are short in the futures, then if price goes up, you are exposed to the additional expense of a margin call. For many producers who become astute observers of the marketplace and will decide for themselves when they want protection and when they don't, futures offers them a great deal more flexibility than does the cash contract.

The third, and only recently available, approach to forward pricing is to use the ***option on futures***. This is an exciting alternative to many producers. It allows protection against falling prices if you are a producer and leaves open the upside price possibilities just in case. We will find later in the book that the flip side of the coin is also there. It was mentioned very briefly in Chapter 4. The processor who uses corn, wheat, or soybeans, for example, or even slaughter hogs, as a raw material is worrying about costs going up over time. By turning around and using the other side of the option picture, the call option, processors can establish a *cost ceiling* on their raw material costs. A big advantage of using the option, we will find, is that in case corn prices come and the cost of that raw material comes down, users will benefit from the lower cost. It is an exciting, new, and very useful addition to the kit of tools in terms of forward pricing possibilities.

All of these approaches, especially the direct use of the futures and the use of options, will be demonstrated and illustrated throughout the remainder of the text. Much of the rest of the book is about helping get the reader to a position of being comfortable with use of all these forward pricing techniques. This way, the individual can pick the one that fits his set of circumstances and preference patterns best and take advantage of what is a very important opportunity. All agricultural producers are price takers on any particular day, but they can do something about the price level they eventually receive if they will monitor pricing opportunities and take advantage of the one thing that they can control: *when a price is set for their product.* To do that, they have to be comfortable with forward pricing techniques.

6

Forward Pricing and Marketing Strategies for Crops

INTRODUCTION

In this chapter, we will look first at the procedure for forward pricing crops using futures and options. We will then turn to look at a storage program and highlight the importance of basis and basis patterns in making that decision. That discussion will be followed by a demonstration of procedures for controlling the exposure to variable prices when you are buying corn, soybeans, or soybean meal as inputs or raw materials. This use of markets has been mentioned in earlier chapters. It is what we call **a long hedge.** We will go through the basic processes and procedures for a growing crop, for a storage program, and for treating crops such as corn as an input to a livestock or dairy program. The latter part of the chapter then turns to examples and case studies for these various decision scenarios.

CASE STUDIES

In the case studies, we will look at comparing cash contracts,with positions directly in futures and options and talk about when each might be used. We will also start helping the decision maker develop and employ a marketing strategy. There has to be some understanding of where information is available and how to determine the probable direction of price movement in making effective decisions on when to forward price.

THE BASIC NEEDS

In the basics for forward pricing a growing crop, there are certain things that the producer needs to have available or on hand:

- **Cost of production estimates and history**
- **Some expectation for the cash-futures basis at the end of the production program**
- **An objective for the marketing strategy or marketing program**
- **Understanding of financing of the program and possible margin needs**
- **Cooperation from the banker**

The basic process will be demonstrated here using wheat and employing the July wheat futures contract as the appropriate new-crop contract. It is important to remember that the futures contract to be used for any growing crop *should always be the contract at or after the end of the growing period* when the cash product can be sold. For example, if we think about a particular type of crop and which futures month turns out to be the widely used new-crop contract, we would have a listing as follows:

Cotton	December
Corn	December
Soybeans	November
Winter Wheat	July
Spring Wheat	December

Note that in wheat we can either be talking about winter wheat or spring wheat, and in winter wheat there are two types. Soft red winter wheat is grown in the Midwest, in the Southeast, and in the Delta states. Hard red winter wheat is grown primarily in the Southern Plains from Nebraska down to Texas. There is also spring wheat, which is planted at essentially the same time that corn is planted, and harvested in the fall. Note the new-crop futures contract for spring wheat is December.

In the U.S. there are three different futures traded for wheat. The Chicago Board of Trade trades the most active contract, and it is essentially soft red winter wheat. Many producers use the Chicago Board of Trade as a place to forward price hard red winter wheat as well, however. Hard red winter wheat is traded on the Kansas City Exchange, specifically the hard red winter wheat from the Southwest and Southern Plains that is of high milling quality for breads. The spring wheat contract is traded on the exchange in Minneapolis.

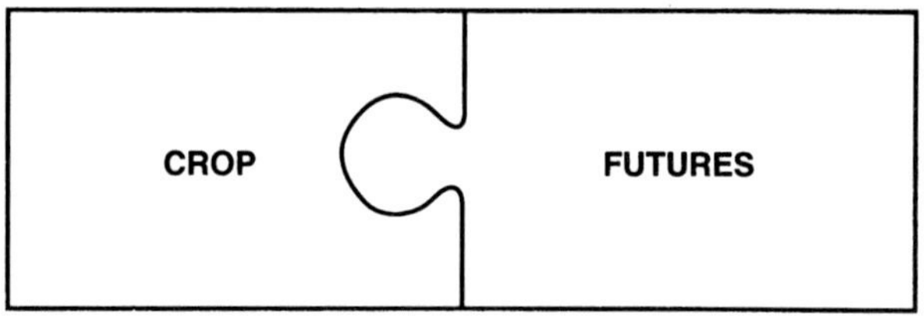

Try to pick the futures that best fits your crop

THE RIGHT FUTURES

The producer needs to understand which futures contracts are most appropriate for his particular type of production. Here,

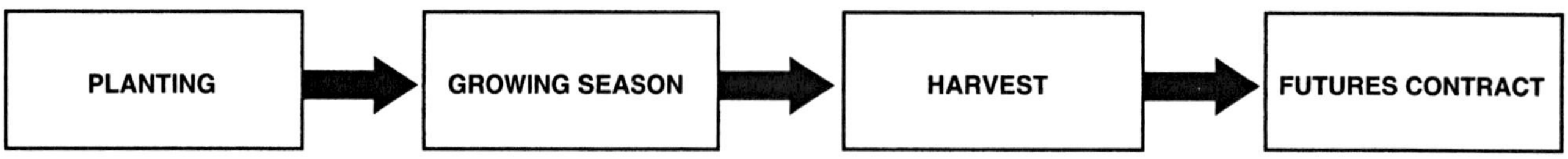

Use the futures contract after the end of the production period

we will assume we are dealing with a producer of soft red winter wheat who uses the July Chicago wheat futures since most harvest of soft red winter wheat is done in June.

IMPORTANCE OF GOOD ESTIMATES

Table 1 provides a listing of the types of information that the decision maker will need to get ready to monitor the markets and prepare for a forward pricing decision. This set of requirements will be there no matter what growing crop is considered. Based on earlier discussion, it should be apparent to the reader that the importance of good estimates of the cost of production and good estimates of the closing basis, the basis at the end of the production program, are especially important. Table 1 shows possible sources of that type of information and goes on to discuss the possible sources of needed financing, plus to the brokerage firms that trade the futures on behalf of the producer.

WRITTEN AGREEMENT

The arrangements with regard to financing of the futures trades and the need for a written three-way agreement between the producer, the banker, and the broker are especially important. The decisions on when forward pricing is conducted should be the producer's after consultation with the banker. The banker needs to agree to the objectives of the program and to provide financing for any margins. If there is limited understanding on the part of any of the participants in this program, there is a need to do some reading, attend short courses, and get better prepared before the program is started. This is especially important with regard to understanding how futures and options are traded "on the margin" and what the financial requirements can be depending on unexpected movements in the market. We will come back to this after a basic hedge on wheat has been explained.

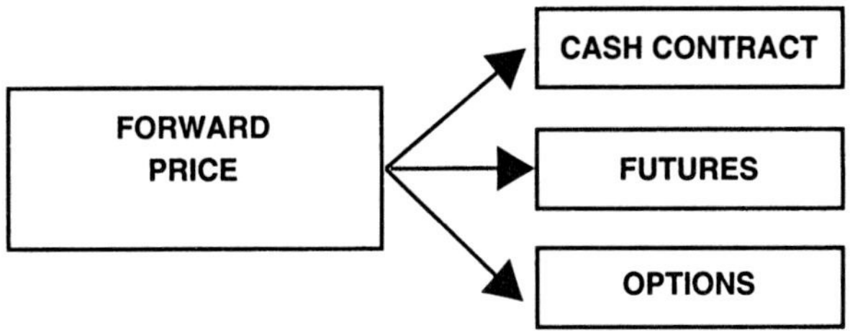

Alternative ways to forward price

Table 2 sets up a demonstration of a forward price for wheat using the futures directly. Keep in mind that something approaching this pricing opportunity would also be available from the local grain elevator via a cash contract. Here, we are assuming that the producer prefers to go into the futures market and do his own forward pricing, perhaps based on the notion that he is, in fact, paying the grain elevator something for the service over time if he uses a cash contract. We will come back later and talk about how this might compare to the cash contract offer and whether or not the producer might prefer to take this approach as compared to using the cash contract

Each part of Table 2 needs some discussion. On November 10 (there is nothing magical about this date), the producer comes up with some of the basic information that he needs. We use

Type of Information	Possible Source
Cost of production	Your own records. Your local extension agent will have a general budget for the crop which you can adapt to fit your particular program if you have no detailed information or format to use.
Closing basis	Your own records or data from your extension agent or university extension specialist. If nothing organized is available, the local library should have microfilm records of past newspapers from which local cash quotes can be collected. Your university specialist, a commodity broker, or your advisory service should have futures quotes. An alternative is the library microfilm of old *Wall Street Journals*. If you have to start from nothing, using the Wednesday cash prices and Wednesday closing prices for July wheat futures will reduce the detail.
Financing	Your banker. There should be a credit line for the price risk management program that is in addition to, and separate from, the production credit line. This is very important. If the bank is reluctant, it is probably best to start with cash contracts or use options on futures, not futures directly
Three-way agreement	This is a *written* agreement that specifies the role and responsibilities of the producer, the bank, and the commodity broker who conducts trades for the producer. *There should be clear procedure to identify who has the authority to make trades.* Normally, this should be the producer in consultation with the banker. The broker typically does not want this responsibility, and his role should be one of placing orders and monitoring the account for margin needs, etc. *The decision is the producer's responsibility*

Table 1 - What you need to prepare for a forward pricing decision

Date	In Cash	In Futures
November 10	Estimate cost of production at $2.40. Based on past records, project cash-futures basis for early June harvest at -$.20 per bushel. Forward Price = Futures + Basis FP = $3.35 - .20 = $3.15 Profit = Forward Price - Costs = $3.15 - 2.40 = $.75 per bushel	July futures trading at $3.35
November 10		Sell July futures at $3.35
June 5	Sell cash at $2.50 Net: $.10	Buy back futures at $2.70 Net: $.65
	Overall Net: $.75 Realized Price = Cash Price + Futures Net = $2.50 + .65 = $3.15	

Table 2-Demonstration of a forward price or short hedge program for a producer of soft red winter wheat

November 10 because the wheat has been planted, is up and growing, and a lot of producers would be reluctant to forward price wheat that hasn't emerged yet. *But there is no reason why this couldn't be done on August 10 before the wheat is planted if the forward pricing opportunities look especially favorable.* Here, we use November 10. It could be a later day in November or any later day toward the harvest time as the producer monitors what type of forward price he is being offered. Let's go back to the information in the table and examine each part of it closely.

CLOSING BASIS

Based on records, the producer estimates the cost of producing wheat at $2.40. This is just the typical cost of production budget for wheat. Very importantly, and based on past records of cash prices *in his local market area*, he expects the cash futures basis for early June at the next harvest to be -$.20. *This says that across the past five years, for example, he has sold wheat in his cash market in early June for around $.20 under the trading level of the July futures in that same early June period.* This latter point is important. On November 10, the producer needs to have information to allow him to project what the cash price in early June will be like compared to the trading level of the July futures in early June. This has nothing to do with cash prices on November 10. The need is for insight into what the cash-futures basis will be the following June. Once that -$.20 basis is established (notice that it is negative), then the forward price is calculated as suggested earlier by adding futures and the estimate of the closing basis. Forward price becomes $3.35 plus the $.20 basis or $3.15. (Actually, of course, it is $3.35 - .20 since the basis is negative.) The profit margin--and this is a new introduction--the producer is looking at is forward price minus cost (the $3.15 forward price minus estimated cost of production of $2.40 or $.75 per bushel).

Once a basis level is selected, it is used all year

PROFIT MARGIN

The producer has to make a decision on November 10. Is this profit margin good enough? It is based on July futures, which, according to the newspaper or based on a telephone call to the broker, are trading around $3.35. Is it likely that the July futures contract will trade still higher? Is it more likely that it will trade lower? All of these things, as we will find later, will roam through the mind of the producer, but there is one final need on November 10. *Is this the day that you will say "yes" to the question of when price will be established?* Or will you continue to be a cash market speculator, carry the risk that prices will go lower, and look for still better prices after November 10? All of this is the difficult part of price risk management, of course. Here, we assume that the producer decides that the offer is good enough, is about as good as it is likely to get, and that he calls the broker and has him or her sell July futures at $3.35.

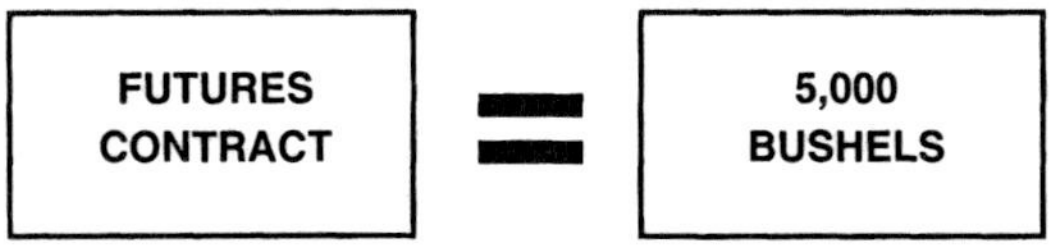

Wheat futures are in 5,000 bushel increments

Notice we have not talked about how many bushels are involved. The entire table is based on a per-bushel basis, so it is correct whether the producer is selling 5,000 bushels, 10,000 bushels, or 150,000 bushels. The wheat contracts traded at the Chicago Board of Trade are in 5,000-bushel increments, so typically it is going to be 5,000, 10,000, or some multiple of 5,000 bushels that the producer trades. So, we assume that on November 10 everything has been prepared in terms of an account being opened with a commodity broker, a three-way agreement has been worked out with the banker, the banker understands the need to get some price protection established at these quite profitable levels, and our producer calls the broker and sells July futures to forward price wheat at $3.35. This first step was not painful at all!

On June 5, again just an example date that suggests we are near harvest, the producer harvests the wheat, hauls it to the local elevator, and has to take the posted price of $2.50 per bushel. He may be upset that wheat prices are at $2.50, but as we have said in prior chapters, there is nothing he can do about it. *He is a price taker, and if he has done nothing to establish price at some other level until he hauls the wheat to town, he is going to have to take that $2.50 price.* He may wait until June 6 and the price may be $2.55 or it might be $2.45, but it is going to be somewhere around that $2.50 level, and he is again a price taker. This is, incidentally, exactly the position that the cash market speculator would be facing. He has produced wheat at a cost of $2.40, he has to sell it at $2.50, and he is only going to net $.10 per bushel. He will probably be disappointed and upset because he is still going to struggle with living expenses, struggle with making annual payments on the mortgage, etc.

THE ROUND TURN

Returning to our example, in Table 2, the producer sells cash wheat at $2.50. On that same day, he calls the broker and is able to buy back the futures that he sold earlier at $2.70. The reader needs to stop and reflect here. Why $2.70? The $2.70 is used in this illustration because we spent a lot of time earlier in the year determining that typically, in early June, cash prices are around $.20 below the trading level of the July futures. *There is no guarantee of this.* A bit later we will demonstrate what some possibilities are, but here we assume that the futures can be bought back exactly $.20 above the selling price in the cash market. Now the producer has a **"round-turn"** completed in the futures market: an earlier sell at $3.35, and a subsequent buy-back at $2.70, which nets $.65 per bushel. If you combine that $.65 with the $.10 in the cash market, we have the overall gross profit of $.75 per bushel. The realized price is always the cash price plus the net from the futures round-turn, in other words $2.50 plus $.65 or $3.15. It is useful to stop and reflect here. The realized price will always be exactly the forward price, and the realized profit will be exactly the planned profit, so long as basis at the close of the program when the cash product is sold and the futures are bought back is exactly the -$.20 that was allowed for in the earlier calculations. Sit back and reflect on this point!

Basis level is not guaranteed

REVIEWING WHAT HAPPENED

Before getting involved in the estimate of what this program would cost the producer (the cost will be small), let's review what has happened. On November 10 when the July futures were sold at $3.35, there is no big amount of money that flows from the producer to the broker or vice versa. What happens is that the producer has an account opened, has necessary margin monies to allow him to trade futures contracts, and on that particular day his account will show a sell or "short" position for X bushels at $3.35. He will usually get telephone confirmation from the broker and will get written confirmation within a few days that this is, in fact, what has occurred. *It is important to keep records and make sure you know what has been done, when it was done, and at what price and quantity level it was done.* Notations on a daily desk calendar that carries all positions forward is often enough. Some electronic market advisory services offer an accounting program that helps you keep track.

WHEN MONEY FLOWS

ROUND TURN COMPLETED

A FLOW OF MONEY DOES OCCUR

There is a flow of money

Now, drop down to June 5. On June 5 when the cash wheat is sold and the producer buys back the futures at $2.70, *there is a flow of money.* There is a profit of $.65 per bushel that has accumulated in the futures account with the broker, and this is now sent to the producer or to his banker (whatever procedure has been set up). Actually, some part of this could have been sent all along as the market dropped from $3.35 to $2.70, but let's assume that it was just allowed to accumulate in the brokerage account. On June 5, the producer will sorely need that $.65 per bushel because he can only sell his cash wheat at $2.50 and generate a very marginal profit of $.10 per bushel. Notice that if the price has been down to $2.30, he would have actually been losing $.10 per bushel, and he would certainly need the net from the round-turn in the futures market. If the cash price is $2.30, then the net from the futures round-turn is $.85 and the realized price is still $3.15! Work on this. What assumption are we making again?

A key point to remember is the price was established at $3.15 after localizing the $3.35 futures quote by adjusting for the -$.20 basis. That particular price level, so long as the basis goes to the -$.20 that was expected, will be exactly what the producer gets. It doesn't matter whether cash wheat prices go to $2.00 or to $4.00. If the market goes up, the producer is going to make a lot of money in the cash market and lose on the round-turn in the futures, yet net the $.75 (before cost of the futures trade)-- and still have a realized price of around $3.15. It is important that the reader understand about when money does and does not flow. The futures are traded on the margin. When you sell wheat futures at $3.35, you do not get $3.35 times the number of bushels you have sold. What you get is an accounting notation. You do not have to have $3.35 times the number of bushels that is sold on deposit to engage in this transaction. What you have to have is roughly 5 percent of the face value of that contract. In other words, a margin requirement has to be met. Later, when you buy back the futures at a lower price, *you do get money sent to you.* If the price had gone up compared to the $3.35, you would have had to answer margin calls and *send more money to the broker to keep the price program in place.* All this will become very clear as we go forward with additional demonstrations. Some detail is needed here to make sure the reader understands the essence of this program.

FORWARD PRICE IN FUTURES	**=**	**SPECIFIC PRICE**

Forward pricing in futures establishes a specific price

GO "SHORT" IN FUTURES

What we have then, from the viewpoint of the producer of a growing crop, is the need to protect against unexpectedly low prices. To do that, since you are growing the crop or what the trade would call being "long" in the cash commodity, you need to sell, or what the trade would call "go short," in the futures market. If all prices come down, your losses in the cash market will be covered by the gains in the futures market. If all the markets go up, your unexpected windfall gains in the cash market will be taken away by the losses in the futures market. *If you forward price in the futures, you are establishing a specific price, and that is essentially what you are going to get when the program is completed.*

In Table 3, we examine some alternative June 5 closes. Two specific alternatives are treated. The first is one in which cash prices go up to $3.25 because of weather problems in the maturing wheat crop during the critical March-April period. This demonstrates how the forward price is still received and how the expected profit per bushel is still received, but it comes from different combinations of nets from the cash and from the futures market.

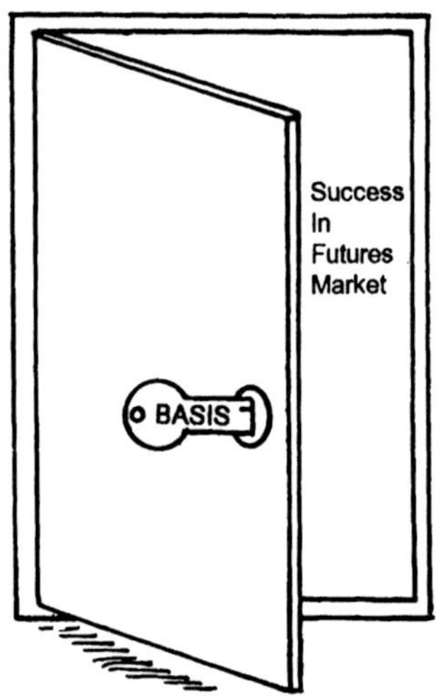

The only reason the specific price will not be exactly received is basis risk

Date	In Cash	In Futures
November 10	Estimate cost of production at $2.40. Based on past records, project cash-futures basis for early June harvest at -$.20 per bushel. Forward Price = Futures + Basis FP = $3.35 - .20 = $3.15 Profit = Forward Price - Costs = $3.15 - 2.40 = $.75 per bushel	July futures trading at $3.35
November 10		Sell July futures at $3.35
June 5	Sell cash at $3.25 Net: $.85 Overall Net: $.75 Realized Price = $3.25 - .10 = $3.15	Buy back futures at $3.45 Net: -.10
June 5 (Alternative)	Sell cash at $2.50 Net: $.10 Overall Net: $.70 Realized Price = $2.50 + .60 = $3.10	Buy back futures at $2.75 Net: $.60

Table 3 - Demonstration of alternative closes to the wheat program

BASIS RISK

The second alternative at the bottom of the table goes back and uses the $2.50 cash price employed in Table 2 but shows that futures have to be bought back not at the expected $2.70 but at $2.75. *What has occurred here is that the closing basis has not been the anticipated and allowed for -$.20 but rather is -$.25. Note that the overall net drops to $.70 per bushel and the realized price drops from the expected $3.15 to $3.10. It is very important that the reader think through this alternative close to the program and reflect on what would happen to the realized price and to the net profits if the futures market does not show the expected -$.20 cash futures basis.* In particular, think about what would happen if the basis is smaller in absolute value than the -$.20 or larger in absolute value than the -$.20. It would be useful here to work through several alternative closes before proceeding. For example, if the closing basis is -.15, can you see why the realized price would be $3.20--no matter what price levels turn out to be?

MARGINS AND MARGIN REQUIREMENTS

Table 4 picks up with the same illustration we have been using. This is important. The producer needs to understand margins and the banker needs to understand margins. It is necessary to take some time to explain each part of this table.

Notice first that there are some assumptions that are made in getting started. First, we assume that the financial position of the producer is satisfactory to the brokerage firm that opened the account. Normally, there needs to be a favorable financial position of the firm with net equity of at least, say, $50,000-100,000 before the brokerage firm will even open an account. This is often true whether you are going to participate in this market as a hedger or as a speculator. We are assuming that all the up-front requirements have been met, and that might include something like a minimum $10,000 balance in this account. When it is not being used, the brokerage firm can put the $10,000 in a money market account (with your permission) and let it earn interest for you. Second, we assume that all paper work has been done, and the account is open. The producer has to sign what is typically called risk disclosure forms to ensure he is willing to accept any losses that occur and is responsible for any losses.

INITIAL MARGIN

We assume that the initial margin requirement is $1,000 per 5,000-bushel wheat contract. *The exchanges on which the futures are traded set minimum margin requirements.* The brokerage firm can ask for more if they can get it and still be competitive with other brokerage firms in the marketplace. What this assumes is that for each 5,000-bushel wheat contract this producer might want to trade to protect his wheat program, the margin requirement is $1,000. Notice that for a 5,000-bushel contract with a price of around $3.50, we are talking about a face value of something in the $17,000-18,000 range. The producer does not have to put up that amount. In fact, he puts up, in this instance, a little over 5 percent of the face value.

Assume: (1)	The financial position of the producer is satisfactory.		
	(2) All paperwork is done and an account is open.		
	(3) Initial major requirement is $1,000 per 5,000 bushel wheat contract.		
	(4) For illustrative purposes, 5,000 bushels are traded.		
Date	**Closing Price**	**Action**	**Account Balance**
November 10	$3.35	Sell @ $3.35	$1,000
November 11	3.38		850
November 12	3.41		700
November 15	3.36		950
November 16	3.31		1,200
November 17	3.27		1,400
November 18	3.16		1,950
November 19	3.20		1,750
November 22	3.32		1,150
November 23	3.40		750
November 24	3.51	Margin Call of $800	200
November 26	3.50		1,050
November 29	3.62	Margin Call of $550	450
November 30	3.62		1,000
December 1	3.57		1,250
December 2	3.58		1,200
December 3	3.51		1,250
December 6	3.44		1,900
December 7	3.46		1,800
December 8	3.39		2,150
December 9	3.35		2,350
December 10	3.30		2,600
December 13	3.32		2,500
December 14	3.30		2,600
December 15	3.25		2,850
December 16	3.20		3,100
December 17	3.22		3,000
.	.		.
.	.		.
June 5	2.70		$5,600

Table 4-Margins and margin requirements for the wheat program

Futures are traded "on the margin"

MAINTENANCE MARGIN

There is a related maintenance margin requirement, and this is very important because it becomes the threshold, as we will see, that triggers a margin requirement. Any time the account balance drops below $600 per 5,000-bushel contract, and we will work with just one contract for simplicity in the table, a margin call is issued. Here, we are assuming a margin is required to take the account back up to the original $1,000 balance. Many brokerage firms, for a hedger, will only require that the balance be taken back up to the $600 maintenance level. That is why it is called a "maintenance level." But the process is exactly the same, and it is useful to assume that you have to take it back up to $1,000 and keep that type of balance in the account. As noted, the fifth assumption is that we will work with a single 5,000-bushel contract for simplicity.

Note that we pick up with the possibility of a closing price, the last price of the day for the July Chicago wheat futures, at

$3.35 on November 10. We assume that on that day, the futures are, in fact, sold at $3.35. There would be a trading range on that particular day, but we assume we will be able to get the contract sold at $3.35. An account balance of $1,000 is associated with this one 5,000-bushel contract. On November 11, the closing price is $3.38. If you had to buy the futures back at $3.38, you would lose $.03 per bushel or $150 on a 5,000-bushel contract. Note the $150 is subtracted from the $1,000 and the account balance now becomes $850. On November 12, the market closes at $3.41, and when the accounting is done at the end of the day, we will find an additional $150 would be lost if you had to buy it back at that closing price, so the account balance is down to $700.

Note the dateline now skips to November 15. This treats the example as if the 13th and 14th were Saturday and Sunday, and, of course, no trade is conducted over the weekend. On the 15th, the closing price is down to $3.36 relative to the $3.35 price at which the position was initially established. You would now be losing only $.01 per bushel, or $50 per contract, if you had to buy it back, so the contract balance is $950. For the rest of that week up through the 19th, the market trades lower, and notice as it trades lower, the account balance grows. For example, on November 19, the closing price is $3.20. If you sold a 5,000-bushel wheat futures at $3.35 and bought it back at $3.20, you would be making on that round-turn action a profit of $.15 per bushel. The $.15 per bushel on a 5,000-bushel contract means an additional $750, so notice the account balance is $1,750. This process is continued, of course, and accounting is done at the end of each day based on the official closing, or settlement price, as it is called at the CBOT.

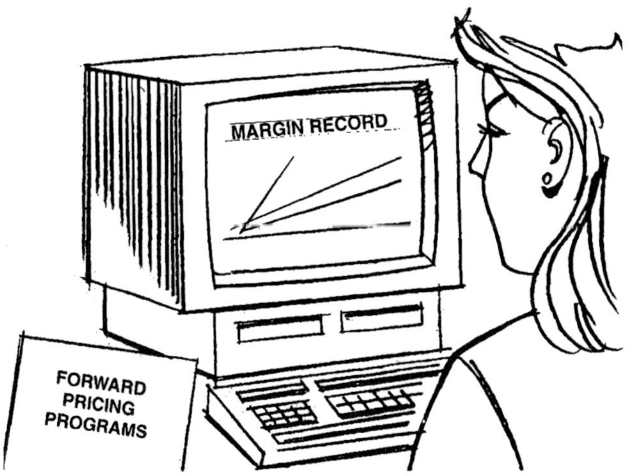

You need to keep up with margins

MARGIN CALL

The next week something happens that starts to push the prices higher, and the situation changes. This would be Thanksgiving week, for example, and notice there is no trade for the 25th, Thanksgiving Day. But by Wednesday before Thanksgiving, something new has happened. The closing price for this contract is up to $3.51. A margin call of $800 is shown. We need to make sure we understand what has happened.

BREAKING DOWN THE BALANCE

If you sold futures at $3.35 and had to buy them back at $3.51, that would mean a loss of $.16 per bushel. That $.16 per bushel on a 5,000-bushel contract is $800. If you subtract $800 from that initial balance of $1,000 (and that balance has worked down toward $750 on November 23), your account balance would be pulled below the $600 threshold. In fact, the balance for that particular day based on the close of $3.51 would be down to $200. That change in the market would necessitate a margin call of $800, which is the amount that would be required to take the account balance back up to $1,000. We treat this as if the margin call has been wired to the brokerage firm and placed in the account. The balance on November 26, the Friday after Thanksgiving, with a closing price of $3.50 now shows $1,050. Let's make sure we understand where that $1,050 balance comes from.

Work with your bank in answering margin calls

At the close of $3.51 on November 24, the account balance was down to $200. A margin call is issued and we assume this is deposited on November 26 in the amount of $800 ($200 plus $800 takes the account balance back up to $1,000). But notice the close on November 26 is $3.50. Relative to the $3.51 that prompted the margin call, we gained $50 on that close. The account balance is up to $1,050.

The next Monday, the market rallies again and closes at $3.62. Relative to where the close was on November 26 with an account balance of $1,050, we have now incurred a $.12 potential loss with the close at $3.62. The $.12 on 5,000 bushels totals $600, and if you subtract that $600 loss from the November 26 account balance of $1,050, you again would come down through the $600 threshold. In fact, your balance on November 29 would be $450. This means a margin call of $550 would be issued on November 29. Note that the amount is what is needed to add to the existing account balance of $450 and take the balance back up to $1,000 again. Thus, on November 30, when the market again closes at $3.62, the account balance has been restored to the $1,000 with the margin call of $550. Things are back in shape again.

You need a separate line of credit to write margin call checks

Starting on December 1, notice that the close is down to $3.57, which puts the account balance up to $1,250. Across December through December 17, the last specific date shown, the market continues to work lower. This would be consistent with improving crop conditions or deteriorating export demand (because the crop is better than expected in some other producing countries). Or any number of supply-demand reasons could be causing this. But the important thing is to notice that by December 17, when the closing price is down to $3.22, the account balance is $3,000. This $3,000 is, of course, made up of the $800 margin call that was sent in on November 24 and the $550 margin call sent in on November 29. That is a total of $1,350. But there is an additional $1,650 that is shown in the $3,000 balance. That comes from the initial $1,000 deposit and the fact that if you sell wheat futures at $3.35 and buy them back at $3.22, there is a gain or profit on that potential round-turn of $.13 per bushel. That $.13 on a 5,000-bushel contract is $650. So to repeat: The $3,000 is made up of the initial $1,000 deposit on November 10, the margin call of $800 on November 24, the margin call of $550 on November 29, and the fact that if you bought a short position established at $3.35 back at $3.22, there would be a gain of $650. The account balance would be up to $3,000.

At this point, it is not very instructive to continue going through each daily close. The table jumps down to the June 5 close that has been used in earlier examples of $2.70. Now, the account balance is up to $5,600. This would be made up, of course, of a $3,250 profit that could be garnered from having sold at $3.35 and bought back at $2.70. That is a profit of $.65 per bushel ($3.35 - $2.70) times 5,000 bushels, or $3,250. The remainder of the $5,600, of course, is made up of the initial $1,000 deposit, which would put it up to $4,250, and the two margin calls, which total $1,350. If you add those all together, you have an account balance of $5,600 on June 5 when the cash wheat is sold somewhere around $2.50 and the round-turn on the futures transaction is completed by buying back futures at $2.70. All this is done through the Chicago Board of Trade as a central clearing house--you do not have to look for someone to sell to or buy from!

WHERE EXCESS MONEY GOES

As the surplus accumulates in the account starting in December and during the year as the market works lower toward $2.70, it doesn't have to stay in the futures account. All you have to have there is the $1,000 requirement or, more specifically, the $600 minimum requirement, which would trigger another margin call. *These excess monies can be transferred into a money market account with the brokerage firm and earn interest, they can be brought back to the bank and credited against the credit line that had been established for the trade in futures, or brought back by the producer and used for various purposes.* Keep in mind that everybody needs to understand how the surplus is going to be handled, and that is especially important for the producer and his banker.

Your banker will be glad to see the surplus funds

What you do *not* want to do as a producer is take this money and spend it on non-business uses. You need to keep in mind that money will be needed to offset losses *or problems* when you have to sell the cash product at or below cost. Not only should the producer be alert to that fact but, of course, the banker should as well. This is something that ought to be worked out in the three-way written agreement between the brokerage firm, the banker, and producer. Specific instructions should be given as to how any excess that accumulates in the futures account is going to be handled.

It is useful to assume that the opposite price move occurs and think about what would happen. Let's assume that instead of seeing the market come down from $3.22 on December 17, the last day shown specifically with an account balance of $3,000, that something happens to the crop in the U.S. during March and April and instead of the $2.70 closing price on June 5, it is $3.70. The reader is encouraged to think about this. Work through and determine what would have occurred as you moved from December 17 toward June 5. Particularly, how much additional margin money would have been required as you move toward a final closing price of $3.70?

BUY BACK FUTURE AT HIGHER PRICES = LOSSES IN FUTURES THAT WILL BE OFFSET BY GAINS IN CASH

If the market climbs after you sold futures, your futures account will show a loss

HIGHER PRICES MEAN LOSSES IN FUTURES

Specifically, if you sold the futures at $3.35 and had to buy them back at $3.70, that constitutes a round-turn loss in the futures account of $.35 per bushel. The $.35 per bushel on a 5,000-bushel contract amounts to $1,650. Since margins in the magnitude of $1,350 had already been sent in back in November, if none of the surplus that had accumulated during December had been pulled out, then only an additional $300 of margin money might have been involved in keeping the account current and properly margined when the futures moved up to $3.70. *The important point to recognize here is that if the market goes unexpectedly higher after it is sold at $3.35*, there will be losses in the futures account. They are paper losses until the final round-turn transaction is completed and the short position in futures is bought back. But once that occurs, then an actual loss has been realized in the form of margin deposits.

THE STORAGE HEDGE

One of the most straightforward, and perhaps easiest uses of the futures market is the storage hedge. The cash-futures basis is employed as an indicator of whether a storable commodity such as soybeans should, in fact, be put into storage or whether it should be sold. What we will find is that it is *very* important to know what both your *opening* and *closing* basis are in evaluating this program. It will again be important to know your costs. It is useful to continue a line of thought in terms of an example. Let's pick up on the wheat example that has already been employed and talk about whether or not this particular crop should be stored. Be reminded that one of the basic illustrations that we have gone through is a scenario in which, at harvest, cash prices were down to $2.50 per bushel. This is down near the cost of production, so there is an immediate tendency to say, "I won't sell this crop. I will store it and count on getting a much better price later on." But that thought process is often faulty. It is important that we look at this decision in an objective fashion.

OPPORTUNITITY COST OF STORING

Table 5 lays out what is needed for an effective decision on storage. On June 6, the producer can sell the wheat at the local elevator for $2.50. *This price is very important because it becomes the value or cost of the wheat going into the storage program.* Notice that the costs of producing the wheat are no longer relevant. Those costs are all "fixed", in economic terminology, and cannot be recovered. What you are giving up is the opportunity cost, the cost you incur by not selling the wheat on June 6 is the $2.50. That is the cost of the wheat in the storage program.

WHAT YOU COULD SELL WHEAT AT IN CASH MARKET = CORRECT "COST" TO CHARGE IF YOU STORE WHEAT

Cost into storage is what a crop would sell for

On that same date, the March Chicago wheat futures are trading at $2.95. If the producer has an estimate of what the cash-futures basis will be in early March and an estimate of the costs of storing for that seven-month period, this decision is relatively easy. Past records indicate that, in early March, the cash wheat prices in this particular local market area have tended to run $.05 under the trading level in early March of the March futures. The basis used as an indication of a closing basis, then, is -$.05. Costs of storing the wheat are estimated at $.28 per bushel, which is about $.04 per bushel per

Date	In Cash	In Futures
June 6	Can sell at $2.50 per bushel Project closing basis for early March to be $.05 Estimate storage costs for June-March at $.28 per bushel Break-even price in early March is $2.50 + .28 = $2.78	March Chicago wheat futures at $2.95
June 6		Sell March futures at $2.95
March 5	Sell cash at $2.65 Net: -$.13	Buy back March futures at $2.70 Net: $.25
	Overall Net: $.12 per bushel	

Table 5-Demonstration of alternative closes to the wheat program

month for the seven-month period. That \$.28 might be made up of costs such as \$.05 per bushel in shrink. This is assessed by using 2 percent, the assumed shrink or weight loss, of \$2.50. The value of that 2 percent shrink and weight loss is \$.05 per bushel. The big cost of the storage program is interest on the \$2.50 wheat that you are holding instead of selling. If you have a note at the bank that you need to renew and extend because you are not selling the wheat on June 6, and that note has an interest cost of 10 percent, then you must calculate 10 percent on \$2.50 for seven months. That turns out to be \$.145 per bushel across the storage period in interest cost alone. There will be some electrical cost in terms of drying the grain to a level at which it can be safely stored and doing some circulating of the wheat during the storage period. Let's assume that cost runs about \$.05 per bushel. Additionally, the producer might want to have some insurance on the value of the wheat in the bins in the event that it gets destroyed by a storm, electrical fire, etc. Let's assume that is about \$.035 per bushel. If you add those up, you get \$.28 per bushel.

COST OF STORING

The \$.28 specific level is not necessarily the most important point here. Your true costs may be different. The point is you need to do a good job of estimating what it costs to hold the wheat and be sure to include an interest charge on the inventory you are maintaining.

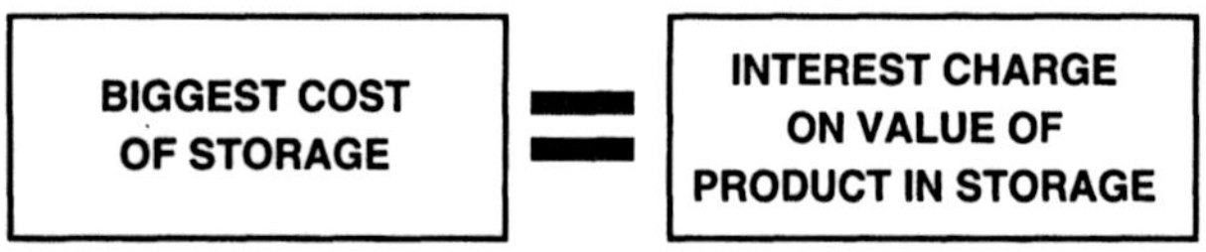

Interest charged on a storage crop is significant

The rest of what is in the table ought to be reasonably easy for the reader at this point in time. The forward price being offered by the March wheat futures, which are trading at \$2.95 on June 6, would be \$2.95 adjusted for that closing basis or \$2.90. The break-even price, if you store the wheat, is going to be the \$2.50 that you are foregoing plus the \$.28 storage cost or \$2.78. If you can forward price the wheat for early March at a price of \$2.90 by selling futures at \$2.95, and the break-even on wheat held until early March is \$2.78, then that leaves a profit margin of \$.12 per bushel.

Let's assume the producer likes this and he does, in fact, on June 6 decide to hold the wheat and sell the March futures at \$2.95. A possible closing of this program is then shown for March 5. Any date in early March would suffice. The cash price might be a bit different, and the price at which you can buy back the March futures would probably also be a little different on different days. Here we assume we sell the cash wheat at \$2.65. That means the price has only gone up \$.15 per bushel, not the \$.28 you needed to have a break-even situation compared to selling back on June 6. So the net in the cash program for the unpriced wheat or the wheat held as a cash market speculator would be -\$.13. You sold March futures earlier at \$2.95. You buy them back at about \$.05 per bushel above the cash price, or \$2.70. Selling at \$2.95 and completing the round-turn by buying back at \$2.70 nets \$.25 in the futures account. If you combine that \$.25 from the round-turn in futures and the \$.13 loss in the cash market, of course you get the expected \$.12 per bushel overall profit.

STORAGE LOSSES

This is a very easy program for the decision maker who is just starting to think about using the futures market. The only risk exposure that is still involved is the possibility that you will lose more than 2 percent of the wheat during the storage period, but good management and good facilities should minimize those losses. The only market-related risk exposure you still have is the possibility that the basis on March 5 will not be the expected -\$.05. If it is -\$.10, then you would end up netting \$.07 per bushel. If it is \$0, then you would net \$.17 per bushel from the program. The reader should make sure he can work through a possible ending that has the cash and futures at first \$.10 apart (that is futures \$.10 above cash) and then have them equal (the futures also being at \$2.65 when cash wheat is \$2.65) to generate those two results. It is a basic program, very easy to apply, and it follows the decision rule that we will develop for storing wheat, corn, soybeans, cotton, or whatever. It is a type of program that the producer can have high confidence will be successful over time.

The storage decision rule

The keys to watch in deciding whether or not to store any commodities such as the wheat discussed in Table 5 is the beginning and ending basis. The decision rule ought to be: *Store so long as the expected improvement in the basis exceeds the cost of storing the grain.*

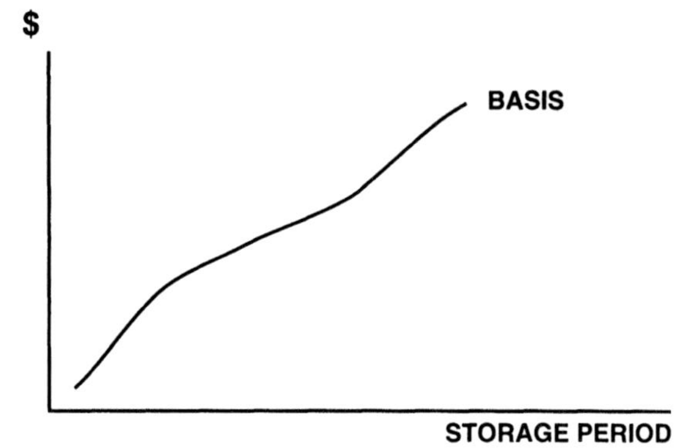

Graph 1 - The increase in the basis is key to profitable storage programs

What this means is that comparing the beginning basis at the time you have to make this decision with the expected closing basis will tell you which you ought to do. *This basis signal is very reliable, and it is a decision guide that the astute producer will want to pick up and use.* Now, let's spend some time on procedure.

The opening basis is the basis on June 6. It is a matter of taking the trading level for the futures contract that you might use to forward price the stored wheat and subtracting it from the cash. In other words on June 6, what we have would indicate an opening basis of -$.45. The March wheat futures are trading exactly $.45 above the $2.50 price level in your local market for cash wheat. The closing basis is the same basis that is used to adjust the $2.95 futures quote down to local conditions. It is the -$.05. What this is saying is that your basis tables show you that in recent years, the cash-futures basis as you move into early March has tended to consistently come to about -$.05. With an opening basis of -$.45 and a closing basis expected to be -$.05, then the expected basis improvement is obviously $.40 per bushel.

STRONG/WEAK BASIS

This move from -$.45 to -$.05 is an improvement because anytime the price of the product that you will sell (the cash) goes up relative to the futures (that you have to buy back), it constitutes an improvement or strengthening in the basis. Typically in the trade terminology, a **"strong basis"** is one that has the cash moving up relative to the futures contract. A **"weak basis"** is one that has cash declining relative to the futures contract. At any rate, what we see here is a projected improvement in the basis, from the viewpoint of the producer, of $.40 per bushel.

STRONG BASIS **=** **CASH PRICE HIGH RELATIVE TO FUTURES**

Basis is strong when you can sell cash relatively high

We have already determined that the cost of storing the grain for the seven months from June to March is $.28 per bushel. If you subtract that cost from the expected basis improvement of $.40, you get the $.12 profit margin that we observed when we did, in fact, hold and forward price by selling futures. If we build on this and elaborate on the decision rule, we can identify some general guidelines as follows:

- **When the basis is unusually weak at harvest, it will usually pay to store the grain.**
- **When the basis is unusually strong at harvest, it will usually pay to sell the product in the cash market and not store the grain.**
- **When the basis at harvest is about at average levels and the basis improvement that might be expected roughly matches the cost of storing the grain, it doesn't matter very much which way you go. Since you extend your exposure to some risk (shrink, basis variability, etc.) by holding the grain, the better decision would be to sell at harvest time.**

The storage decision is not all that difficult

STORAGE PROFITS NOT GUARANTEED

There are several interesting implications to these rules. Just because the cash price is low at harvest, lower than you would like to see, in no way ensures that a storage program will be profitable. Cash prices will be low when the supply of grain is excessive. *If the supply of grain is excessive, the marketplace does not have to pay you something above the cost of storage to encourage someone to hold wheat from June until the following March.* In fact, there are likely to be so many people wanting to store because they don't like the idea of selling at $2.50, that we shift wheat out of the June period and into the March period. This shift drives the March price down relative to June and *makes sure the storage program will not be profitable!* You have to be very careful adopting the argument that if selling prices at harvest are lower than you would like to see, that it makes sense to store the product. What *does* make sense is to look at the beginning basis at harvest, calculate what the improvement would be if the market goes to the closing basis that you have seen in past years, and then make your decision.

WHAT ABOUT THE EMPTY BINS?

Some producers will be bothered by the idea that in years in which the basis is favorable or strong at harvest, you shouldn't store the grain. The immediate reaction will be, "But I have built on-farm bins. What am I going to do with them?" The appropriate answer here is "almost anything, just don't insist on losing money by storing the wheat." Keep in mind that the cost of putting up those bins was incurred in some past years. It is fixed. Whether you use them or not, they are going to be there. There is nothing you can do to change that. Bin capacity ought to be used in years in which the basis patterns offer some encouragement for storage, and the bins should sit empty during those years in which the basis is very strong or very favorable at harvest time. A strong basis at harvest is the market's way of telling you that it wants your grain then and not to store it.

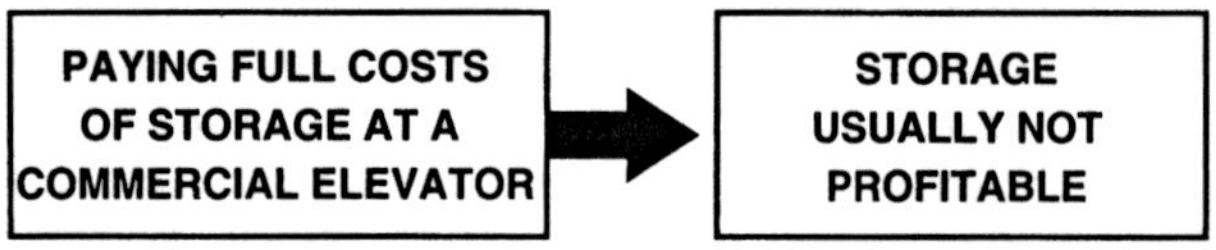

It costs more to buy space

This discussion would also be valid for a producer who does not have storage on the farm but might be interested in buying capacity from the local grain elevator. Often, the grain elevator manager will offer various types of programs to encourage you to rent their space. They might hold the commodity from June to September, for example, at a cost of $.10 per bushel, and then it will be $.05 per bushel per month from September forward. You have to back away from these offers and assure they make sense. Go back to the indicators that would be coming from the basis patterns and use those to determine whether or not you should store the grain. As a general rule, the variable costs or out-of-pocket costs associated with storing in on-farm bins is going to be lower than what you will have to pay the local grain elevator to rent their space. There will be fewer years, therefore, in which you ought to store if you have to rent the capacity as compared to already having an investment in on-farm bins.

DON'T STORE AS COST MARKET SPECULATOR

If you go back and examine what has happened historically for corn, wheat, soybeans, cotton, etc., in any 10-year period there will typically be two to four years in which storage made no sense at all. Anybody who stored the commodity as a cash market speculator and didn't forward price it really got hurt. Storage costs were substantially above the price increase from harvest until later in the year. There will typically be maybe four to six years in which it didn't make much difference which way you went. In other words, the basis improvement and the increase in price in the cash market was about consistent with the cost of storage. Therefore, which decision you made in those years didn't matter very much. *There will typically be one to two years in which having access to storage space on the farm was extremely profitable.* Those are years in which the basis is unusually weak at harvest time, and you can take advantage of your knowledge that it is likely to go to the -$.05 demonstrated in Table 5. You can store the product and profit from it. Clearly, having good basis data is useful and important!

You need to look at frequency of storage profits

YOU HAVE TO FORWARD PRICE

You must keep in mind that looking at the basis patterns and projecting how much you expect basis to improve doesn't do any good if the product is not forward priced. What we are saying, then, is that corn, wheat, or soybeans ought to be held in storage only when the projected basis improvement is sufficient to more than cover the cost of storage. Translated, that means that at harvest time you can forward price the corn, wheat, or soybeans in futures at a price above the break-even calculated from what you could sell the cash product for plus the cost of storage. *But none of this works in terms of using the basis as a guide if you don't proceed to forward price the product.* It might even be done with a cash contract. But keep in mind that just because you expect the basis improvement to be, say, $.40 as in Table 5 and you hold the wheat as a cash market speculator, there is nothing to keep the entire market complex from moving lower under the weight of increased wheat production in some other producing country. The storage program then becomes a losing proposition. *If you are going to use the cash-futures basis as an indicator as to when you ought to store, then the rule is store only when you can forward price at a profit.* In those years in which it appears that you can't, the best thing to do is to sell the commodity.

If they really think wheat, corn, soybeans, or whatever is going to move up over time, what a lot of people will do is buy the futures and hold that position instead of holding wheat. This has some interesting ramifications in terms of whether or not you are a hedger or a speculator in futures, and it may raise some tax ramifications. You ought to always talk to your accountant before undergoing any such program. The big advantage of "holding the paper," of course, is that it eliminates storage costs!

LONG HEDGES ON INPUTS

We have seen possible uses of the futures markets demonstrated on a growing crop and on a storage program. Both of these decision situations deal with a producer or a holder of inventories. In every case, the appropriate action in the futures market was to sell the futures. That is because the concern is that price will go down during the growing season or during the storage program. Selling the futures, or going short as the trade calls it, protects you because as the cash market goes down the futures market will go down too. Later in the chapter we will look at some case examples extend the use of these markets to options, and compare with cash contracts, etc. But the intent now is to cover the basic decision areas that the producer or user of grains and fiber crops would face.

The third very important use of the markets is to control the cost of inputs. This is a *long hedge* decision scenario. What we have is a livestock producer, a poultry firm, a dairy farmer, or someone who is buying corn, soybeans to process, or soybean meal as a protein supplement and is worried about the cost of those important inputs going up. The need is to be able to employ some strategy that will protect against rising cost of inputs, and this is where the long hedge comes in.

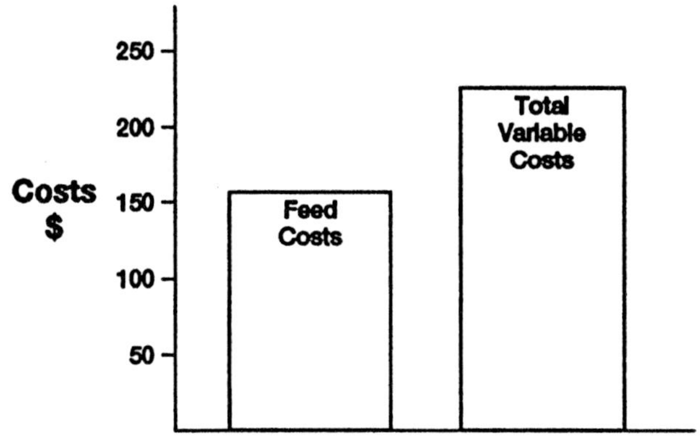

Graph 2 - Long hedges help keep input costs down

CATTLE FEEDING

Table 6 frames out a long hedge for a cattle feeder who will be needing to buy corn during the summer and early fall months to feed 700-lb. steers that are expected to reach market-ready weights of 1,100 lbs. by late October.

Basis in long hedges

The important ingredients in this table ought to be familiar to

Date	In Cash	In Futures
May 5	Placing 700-lb. steers in a feedlot and project them to weigh 1,100 lbs by late October. Cash-futures basis June =+$.05 Cash-futures basis August =-$.05 Cash-futures basis October =-$.20 Cost of corn for late May, June = FUTURES + BASIS = $2.60 + .05 = $2.65 Cost of corn for late July, =$2.55 - .05 =$2.50	July corn $2.60, September corn $2.55, December corn $2.50
May 5		Buy July, September, December corn futures at $2.60, $2.55, $2.50 respectively
June 15	Buy cash corn at $2.90 Net cost of corn = Cash + Net from futures = $2.90 - .25 = $2.65	Sell July futures at $2.85
August 20	Buy cash corn at $3.30 Net Cost = $3.30 - .80 = $2.50	Sell September futures at $3.35

Table 6 - A long hedge on corn for a cattle feeder

the reader by now. Again, basis is important. Notice that on May 5, the July corn futures are trading around $2.60, the September around $2.55, and the December around $2.50. The cattle feeder may well be concerned that weather problems during the summer months this particular year could drive corn prices higher. He needs some of those same basic data that have been employed in other decision situations. From past records, he knows that in his particular area, cash corn is typically selling $.05 per bushel above the July futures in June. By the time he moves out toward August, using the September futures, past records show the basis to be near -$.05. And then in October, the basis is -$.20, as cash prices tend to run $.20 below the trading level of the December futures.

This basis pattern would be fairly typical of any region in the country, where cattle feeding might be going on, that is some distance away from the center of corn production. If you go to central Iowa, for example, you would seldom find a positive basis such as that shown during June as the cash price during June is compared with the July futures. But this could happen in southwest Kansas where a lot of the cattle feeding activity is located and costs of corn get bid up. Given the basis information and given the trading levels of the various futures contracts that will be employed, our cattle feeder can now determine what corn is going to cost.

Basis is also important in long hedges

Note that the cost of corn for late May and early June, for example, is futures (July futures) plus the basis. That means $2.60 plus +$.05 or $2.65. The cost of corn for late July would be somewhat different. Late July and August would use the September futures, and here we have $2.55 minus the $.05 since basis is negative, or $2.50. The reader might want to make sure that he understands why the cost of corn that will be purchased, say, in late September or October using the December futures would be $2.30. Make sure you know how to take that $2.50 trading level for December corn futures, adjust it for basis for that October time period, and get down to a cost of $2.30. Remember, the relevant basis expectation is -$.20!

A POSSILBLE ENDING

Table 6 shows only one possible ending. On June 15, for example, we assume that some worries are starting to emerge about a late planted crop not giving the yields that were expected. The cash corn is at $2.90, and the July futures are $.05 below that as had been expected, trading around $2.85. So the net cost of the corn that has to be bought in June, assume June 15, is the cash price plus the net from the futures. This is the $2.90 cost minus $.25, or $2.65 per bushel. Go back and make sure you know where the $.25 from the round-turn in futures comes from. Remember, you *bought* July corn futures at $2.60. If you can sell them at $2.85 on June 15, your round-turn before commission costs, etc., will net you $.25 per bushel. You take that $.25 and subtract it from the $2.90 you have to pay to get your corn, and you are back at $2.65. So, before commissions and any interest charge on margin money, the corn costs a net of $2.65 per bushel.

The next time corn has to be bought is on August 20, and the cattle feeder's worst fears have been realized. Weather has been bad during the pollination period for corn in July, and there are widespread expectations that the corn crop will be 20-25 percent below earlier expectations. Cash corn is up to $3.30 per bushel, and the cattle feeder is glad that he has, in fact, a long hedge program in place. Notice that he will now be able to sell September futures at $3.35, assuming that the -$.05 basis does materialize. If he subtracts the $.80 round-turn net in the futures (buy at $2.55, sell at $3.35) from the cash cost of $3.30, he is back to the expected $2.50 per bushel. If the next time the corn is bought is on October 12 (and the December futures would have been used), and if cash corn is at $3.10, the cost of corn is going to be in net $2.30, assuming that the -$.20 closing basis does, in fact, develop. Work through that and make sure you appreciate the importance of this type of program to protect against rising input costs before moving on.

LONG HEDGES TO KEEP COSTS UNDER CONTROL	=	PROFITABLE FARM PROGRAMS MORE OFTEN

Opportunities in reducing costs are often greater than opportunities on the selling side

EXAMPLES, CASE STUDIES: CROPS

The procedures are much the same for all crops. Whether during the growing period or when considering storage, the concern is the same: a sharp decline in price could be disastrous. The producer who is coming to understand the importance of marketing will want to look at getting some type of price protection. To do that will require comfort with a cash contract for future delivery, a futures position, or an option. Let's lay out some scenarios to demonstrate which might be the correct approach and follow through with a decision in each case.

CASE 1: PRICING SOYBEANS

It is May 20. Recent rains have delayed planting of soybeans, and futures prices are increasing. You expect to produce 50,000 bushels of soybeans and recognize that, financially speaking, it will be difficult to absorb the financial pain if soybeans go to $5.50 or lower. In response to an earlier request, the extension specialist at the state university sent you yearly average soybean prices since the mid-1970's. You have them in a computerized spreadsheet and have been updating the data each year. You pull up the table and find the information listed in Table 7.

Crop Year	Production (mil. bu.)	Ending Stocks (mil. bu.)	Average Price ($/bu.)
1970-71	1,127	99	2.85
1971-72	1,176	72	3.03
1972-73	1,271	60	4.37
1973-74	1,548	171	5.68
1974-75	1,216	188	6.64
1975-76	1,549	245	4.92
1976-77	1,289	103	6.81
1977-78	1,767	161	5.88
1978-79	1,869	176	6.66
1979-80	2,261	358	6.28
1980-81	1,798	313	7.57
1981-82	1,989	254	6.07
1982-83	2,190	345	5.71
1983-84	1,636	176	7.83
1984-85	1,861	316	5.84
1985-86	2,099	536	5.05
1986-87	1,943	436	4.78
1987-88	1,938	302	5.88
1988-89	1,549	182	7.42
1989-90	1,924	239	5.69
1990-91	1,926	329	5.74
1991-92	1,987	278	5.60
1992-93	2,188	292	5.56
1993-94	1,809	155	6.45

Table 7-Soybean production, ending stocks, and average price, 1970-94

As you work on your marketing history you'll increasingly see the importance of big crops and of big and burdensome ending or leftover stocks. In the mid-1980's, you will see big crops, big stocks, and the only average price below $5.00 since 1975-76.

The micro-macro trap?

Here, you step back and think. In late March, the USDA released a Prospective Plantings report that suggested producers will plant over 60 million acres--a number you know is up compared to recent years. Some ominous thoughts start to creep in--will we see the "micro-macro" trap this year? It's time to look closely at the forward pricing opportunities.

Get needed information to compare your alternatives

Some needed information:

- **A phone call to the grain elevator indicates the cash contract offer for October 15-30 delivery is $6.55.**
- **The electronic market wire service in your office shows November soybean futures trading near $6.95. You note the life-of-contract high for the November that started trading months ago is $7.01 and the life-of-contract low is $5.74.**
- **Past records on October basis (cash vs. November futures) show an average basis of -$.28 and a range of -$.46 to -$.14 across the past five years.**
- **The premium on a $7.00 put option (strike prices for soybeans are in $.25 intervals) on November futures is trading near $.23 per bushel.**

You focus on:

- **The cash contract offer of $6.55 is reflecting a basis of -$.40.**
- **Basis has been as weak as -$.46 across the past five Octobers.**
- **The average basis of -$.28 suggests a futures forward price of $6.95 - .28 = $6.67, a $.12 better price than the cash contract.**
- **The $6.95 futures price is near the all time high for the contract. Using a $7.00 option gives you up-side potential, but the market will have to make new highs if the advantages of using put options are to be realized.**
- **Your banker is fairly new at all this, and he is not sure the bank's credit committee will approve anything more than a small credit line to handle margin calls. You know you cannot handle the margin calls if major weather problems develop and November futures go to $7.50 or $8.00.**

When all this is pulled together, what is the correct decision? Should you forward price now? If "yes," which approach should you use?

The decision

A reasonable decision is to take the cash contract. There are good reasons to take these favorable and profitable prices: Your costs of production are near $5.25 per bushel. Odds favor a return to normal weather, a big crop from the increased acreage, and lower prices later in the year. It is too soon in your ongoing development as a marketer to use futures--the financing is too shaky. Even the added $.12 from going directly to futures is not enough, perhaps, because the closing basis in October might be -$.40 or weaker, not -$.28. There is no convincing reason to pay the premium for the put option--you expect prices to go down, not up. The premium is for upside flexibility that is not likely. The decision: go to the cash contract and get protection.

THE CASH CONTRACT IS A GOOD CHOICE HERE

Taking the cash contract eliminates any margin call problems

CASE 2: STORING WHEAT

It is June 25. You have harvested wheat and put it in on-farm storage while you completed harvest, which also allows you to stay away from long lines of trucks at the local grain elevator. You now sit down to deal with the possibility of storing the wheat until the following March.

"YES" ON THE STORAGE DECISION → **WHEN** → **OPENING BASIS IS WEAK**

The basis patterns tell you when to consider storing

You have this information:

- **Bids at the elevator are at $2.90 per bushel.**
- **The March futures are trading at $3.40, making the current or "opening basis" -$.50 ($2.90-3.40).**
- **Your basis records suggest the basis near March 1 is usually near -$.15, which means the "closing basis" estimate will be -$.15.**
- **The premium on a March put option for a $3.40 strike price is $.08 per bushel.**
- **Based on past records and adjusting for the interest cost on holding wheat that you could sell at $2.90, you estimate your costs of storage June-March at $.25 per bushel.**
- **Cash contract bids for early March delivery are at $3.20.**
- **The crop is assured in the U.S., but there are indications of crop problems in other parts of the world that will be harvesting later in the calendar year.**
- **The life-of-contract price range for the March futures is $3.05 to $3.87.**

You reflect on what you have:

- **The cash contract for early March will mean a profit of $.05 per bushel given that your early March break-even is $2.90 + .25 = $3.15. (You do remember that the correct cost of putting the wheat into a storage program is the $2.90 you do not get if you store.)**
- **The forward price in the futures is $3.25 (3.40 - .15), but you know the basis will not necessarily be -$.15 at the close.**
- **A forward price of $3.25 means a profit of $.10 per bushel.**
- **The floor price set by the option will be $3.17 ($3.40 - .15 - .08), which assures essentially a break-even if prices go down, since $3.15 is needed to break even in early March compared to selling at $2.90. You are in a reasonably sound financial position and will not have to borrow money to allow you to store the wheat.**

Do you store? If so, do you forward price? What instrument or approach do you use?

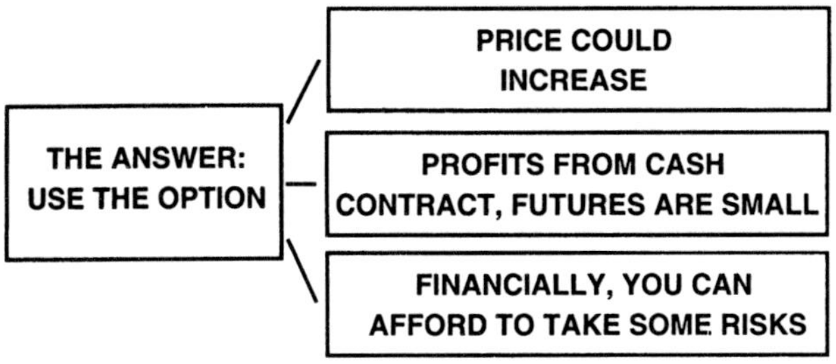

Options fit here for a number of reasons

The decision

A reasonable decision is to store the wheat and buy the $3.40 put option. Profits will be small if you take either the cash contract offer or sell March futures near $3.40. You are in business to make profits, and there appear to be some reasons to expect higher wheat prices later in the year, given emerging problems in other producing countries. Buying a $3.40 put essentially ensures you will do no worse than selling now, and it does leave open the possibility of benefiting from price increases later after the supply pressures at harvest subside.

An effective forward pricing program has great potential to boost profits

SUMMARY

There are a number of alternatives that allow the crop producer to get away from the status of being a price taker. You cannot control price to ensure a profit, but can control when you price. This chapter demonstrated and discussed the different approaches to price risk management via forward pricing alternatives.

To do it right, you need to meet some important up-front requirements:

- **Know your costs**
- **Know your financial position. If your financial position is weak or your banker is not yet ready to provide margin financing, you probably should not go into the futures directly.**
- **Basis information is critical. Estimates of the closing basis allow you to know what forward price you are being offered and to analyze the competitiveness of cash contract offers. For storage decisions, you have to know the opening and closing basis levels.**
- **Which approach is right will depend on financing, your own ability to carry exposure to price risk, the probable direction of price trends, and where you are in the price range for the year.**
- **When using put options, the two big differences from positions directly in the futures are the lack of exposure to margin calls and the opportunity to benefit from unexpected price increases.**

Managing you're exposure to price risk need not be difficult. Knowing the alternatives and then being willing to act on acceptable prices--especially when they offer attractive profits--is the key. Discipline is needed, and it is always important to step back, watch for the micro-macro trap, and protect yourself from a price disaster.

7

Forward Pricing and Marketing Strategies for Livestock

INTRODUCTION

Chapter 6 developed the concept of a hedge or a forward price for crops. Many of the developments in that chapter will be appropriate again as we deal with livestock. For example, there was considerable attention paid to margins and margin requirements in Chapter 6. Table 4 of that chapter provided detailed data on closing prices for a wheat contract and followed through with demonstrations of margin calls and what the account balances would be, etc. Although we are now talking about livestock, the procedure and the process is very much the same.

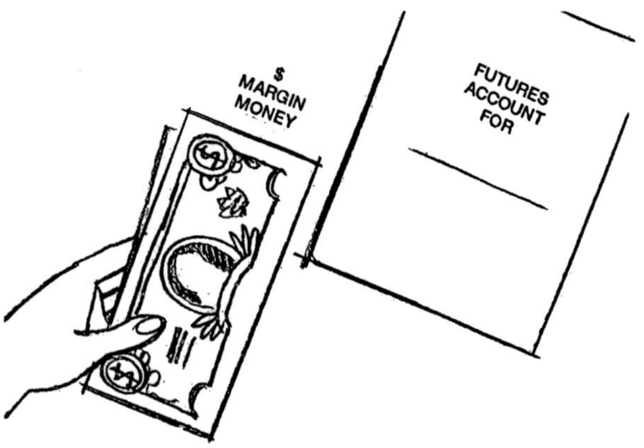

The margin needs are there for livestock programs as well

Recognizing that there will always be readers who will turn first to the chapter on livestock, however, the intent here is to provide essentially a complete coverage. Reference will occasionally be made to specific examples or illustrations in Chapter 6 to avoid needless duplication, but the coverage for livestock will be complete and self-contained to meet the needs of the reader who has particular interest in the livestock commodities. We will not, of course, need to deal with a storage program here. But the long hedge programs introduced in Chapter 6 are very important to a livestock producer. The possibilities of using cash contracts, futures directly, or other types of contracts tied to futures are important and will also need to be covered.

BASIC INFORMATION

When forward pricing a livestock program, there are certain pieces of information that have to be available.

- **Cost of producing the cattle or hogs**
- **Historical data on which to estimate the cash-futures basis or the closing basis at the end of the production program**
- **Understanding of the financing of the program and possible margin needs**
- **An objective for the marketing strategy or marketing program**
- **Cooperation from the banker**

Many of these are the same as the list presented for crops in Chapter 6, of course. The same basic pieces of information will always be required no matter what the commodity might be.

THE LIVESTOCK MARKETS

Table 1 shows the months for which cattle, hogs, and feeder cattle contracts are traded. The live cattle contract is the fed cattle or slaughter cattle contract dealing with steers and heifers coming out of the feedlots weighing from 1,000 to 1,300 lbs. That contract is basically traded in alternate months starting with February, then April, etc. The live hog contract trades for a similar set of months, except July is added. This is the contract for slaughter hogs in the 230 to 250-lb. range. July is a seventh trading month for live hogs.

	Months											
Contract	**J**	**F**	**M**	**A**	**M**	**J**	**J**	**A**	**S**	**O**	**N**	**D**
Live Cattle		X		X		X		X		X		X
Live Hogs		X		X		X	X	X		X		X
Feeder Cattle	X		X	X	X			X	X	X	X	

Table 1 - Trading months for cattle, hog, and feeder cattle futures contracts

The feeder cattle contract shows a substantially different availability. Historically, emphasis has been on sales of cattle in the spring months--March, April, and May--and in the late summer and fall months--September, October, and November. This offering of contracts was designed to fit the production patterns. Light cattle, called *stocker cattle* when they go into grazing or forage programs, are either carried through the winter to be sold in the spring months, or they are carried during the grazing season to be sold in the fall months. Thus, the clustered offerings of contracts were in the spring and fall months. A few years back, there was considerable criticism of the long vacant time span between November and March. A January feeder cattle contract was added by the Chicago Mercantile Exchange, and it offers an alternative month. The January contract has never traded as actively as the others, but it continues to be viable and continues to be offered.

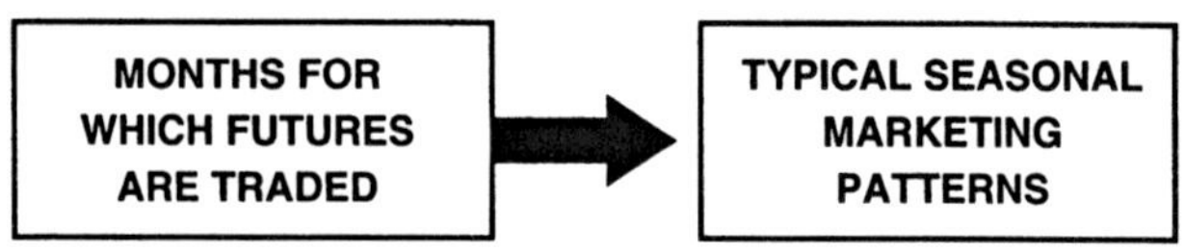

Futures months differ for cattle and hogs to fit needs

THE TRUCKLOAD EFFECT

The sizes of the contracts have changed over time. Currently, the contracts call for 40,000 lbs. liveweight for both the live cattle and live hog contracts and 50,000 lbs. for the feeder cattle contract. It may be that the size of the live hog and live cattle contracts will be increased in the near future to the 50,000-lb. level, which approximates a truckload. Many of these livestock are moved by trucks over the highway systems in the U.S. It is important to keep in mind the size of the contract in determining how many contracts will be needed to provide a certain level of coverage. Depending on the precise weight, of course, each live cattle contract would amount to 35 to 38 finished steers. For a cattle feeder who may be feeding 10,000, 20,000, or even 50,000 head of cattle at one point in time, a number of contracts would clearly be needed. The same point can be made about the hog contract.

FEEDER CATTLE CONTRACT MAY BE TOO BIG

The difficulties with the feeder cattle contract have tended to run in the other direction. The average beef cow herd in the United States is between 40 and 50 head. Such a relatively small producer would not produce the 60 to 70 steers or heifers weighing 700 to 800 lbs. that would be needed to constitute even one feeder cattle futures contract. There have been suggestions over time that this effectively means that the very small producer cannot use the markets. The criticism has some validity. The bulk of the use, of course, is by larger stocker operators who either have many multiples of the number of cattle in the feeder cattle futures contract to sell or by the feedlots who may buy feeder cattle in the thousands. The contract size is no problem for these larger sellers or buyers.

The feeder cattle futures are too large for some producers

BASIS--AGAIN

The initial discussion of a hedge in livestock will employ the feeder cattle futures. This need probably extends across a much larger group of readers than would the live cattle contract or possibly even the hog contract. Table 2 is an adaption of Table 1 in Chapter 6 that showed basic types of information and possible sources of that information when discussing a pricing program for wheat. The table has been adapted to the needs of a feeder cattle program. The reader will note that many of the components are very similar to those present when the decision process involved a storable commodity like wheat. Table 2 is prepared on the assumption that we are dealing with a grazing program in which the steers will be sold in late September or early October, which means we will be using the October feeder cattle futures.

The issue of basis needs extended discussion. Experience has shown that for all of the agricultural commodities that might be forward priced using futures, the basis for feeder cattle may be the most variable or least predictable. A moment's reflection suggests why basis is a bit more complex with feed-

Type of Information	Possible Source
Cost of production	Your own records. Your local extension agent will have a general budget for summer grazing programs or wintering programs which you can adapt to fit your particular program if you have no detailed information.
Closing basis	Your own records or data from your extension agent or university extension specialist. If nothing organized is available, the local library should have microfilm records of past newspapers from which local cash quotes can be collected. Your university specialist, a commodity broker, or advisory service should have futures quotes. An alternative is the library microfilm of old *Wall Street Journals*. If you have to start from nothing, using the Wednesday cash prices and Wednesday closing prices for October feeder cattle futures will reduce the detail.
Financing	Your banker. There should be a credit line for the price risk management program that is in addition to, and separate from, the production credit line. This is very important. If the bank is reluctant, it is probably best to start with cash contracts or use options on futures, not futures directly.
Three-way agreement	This is a *written* agreement that specifies the role and responsibilities of the producer, the bank, and the commodity broker who conducts trades for the producer. *There should be clear procedure to identify who has the authority to make trades.* Normally, this should be the producer in consultation with the banker. The broker typically does not want this responsibility, and his/her role should be one of placing orders and monitoring the account for margin needs, etc. *The decision is the producer's responsibility.*

Table 2 - What you need to prepare for a forward pricing decision for feeder cattle er cattle. The cattle going into the grazing program that we are thinking about here during, say, late February or March may weigh 450-500 lbs. Those cattle will grow, of course, and how much gain there will be is unpredictable. It depends on the weather and a host of other factors beyond the producer's control. A reasonable projection might be that they will weight 800 lbs. in late September or early October. The need then is for an expectation of what the cash-futures basis will be in late September and early October for 800-lb. steers.

Basis in feeder cattle varies with grade and size of the cattle

The feeder cattle futures contract is written for a certain quality of feeder cattle. Details on the contract specifications are available from any commodity broker or the Chicago Mercantile Exchange and need not be reproduced here. But it is important to recognize that there is a quality factor involved. Contract specifications call for M1 or some number of L1 steers to be a "correct" unit. This terminology refers to a medium frame (the "M") #1 muscled steer and a large frame (the "L") #1 muscled steer respectively. If your cattle are projected to be M2, which means medium frame, #2 muscled (thinner muscled) cattle, then the cash price would likely be somewhat lower. The cash-futures basis that would need to be used for M2 cattle, then, will be somewhat different from the basis that should be used for M1 cattle.

The need here is to talk to local extension personnel, market operators, and people who might be conducting special sales for the state feeder cattlemen's association, and get some indication of the normal price differential between M1 and M2 steers weighing 800 lbs. That way, when you make the adjustment of the futures price down to your local market conditions, it will be the proper and correct adjustment, and there will be no surprises at the end of the program. To repeat an earlier observation: Even if you have the right quality of cattle, basis for feeder cattle will still be relatively variable and you need to recognize that.

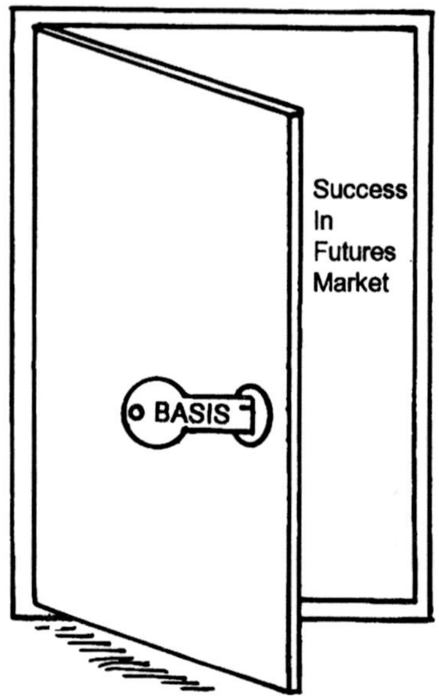

Basis variability in feeder cattle hurts the effectiveness of hedges

Feeder cattle basis is variable

Researchers have not been able to identify the precise reason why feeder cattle basis appears to be rather variable, but much of it may be wrapped up in the slowness with which cash markets around the country adjust to each other. The feeder cattle futures price that is discovered at the Chicago Mercantile Exchange was, in early 1995 (it has changed over time), based on steers weighing 700 to 800 lbs. in 12 midwestern states. It may take some time for a price change in the cash market in Indiana, for example, to filter through the system and be reflected in the cash market in the southern half of Georgia. Without getting too bogged down in detail, there are many reasons why the basis across the country could be variable, and the user does need to be aware of the situation.

AN EXAMPLE

Table 3 provides a hedging example. The top part of the table lays out a total cost estimate of $630, or $78.75 per hundredweight, to produce an 800-lb. steer. Note that these costs are made up of the initial cost of the 450-lb. steer calves and an estimate of $180 as the cost to carry each animal on grass into the fall months. This $180 per head would need to cover an allowance for death loss, vet and health expenses, interest on the money tied up in the steer calves that were bought, labor, salt, minerals, any supplemental feeding, and any other variable cost present because the cattle program is being conducted. Most land grant universities and local extension personnel will have budgets for this type of program. All this results in a total cost of $630 which means a break-even on the 800-lb. cattle of $78.75 per hundredweight ($630 ÷ 8.0 cwt.).

On March 17 when these cattle are bought, the October feeder cattle futures are trading at $85.00 per hundredweight. March feeder cattle futures are offering at least a modest

Date	In Cash	In Futures
March 17	Buy 450-lb. steer calves at in-pasture cost of $100 per cwt. or $450 per head. Estimate 350 lbs. of gain at a cost of $51.50 per cwt. or $180 per head. Total cost will then be $630 per head or $78.75 per cwt. Estimate closing basis at -$2.00 per cwt. Forward Price = Futures + Basis FP = $85.00 - 2.00 = $83.00 Profit = Forward Price - Costs = $83.00 - 78.75 = $4.25 per cwt.	October feeder cattle futures . trading at $85.00 per cwt
March 17		Sell October feeder cattle futures at $85.00 per cwt.
September 25	Sell 760-lb. steers at $76.00 per cwt. Net: -$2.75 Overall Net: = $5.25 per cwt. Realized Price = Cash Price + Futures Net = $76.00 + $8.00 = $84.00	Buy back October feeder cattle futures at $77.00 per cwt. Net: +$8.00

Table 3 - Demonstration of a short hedge on feeder cattle to be sold in the fall months

profit on these cattle so the producer embarks on this program. Assume he has good records on what it costs to produce cattle and has done a good job of adapting state university budgets to his particular grazing situation. He should therefore be able to estimate the cost of growing these cattle--assuming anything close to normal weather and grazing conditions--with a fairly high degree of accuracy. *You can go through all of this on paper and recognize that a profit is being offered even before the cattle are bought. This helps you know how much you can pay for the calves--a very important decision!*

Basis adjustments

At a trading level of $85, after adjusting for the -$2.00 basis, the forward price being offered for the cattle is $83. The reader should recognize that if we were dealing with 700-lb. cattle to be sold in late September or early October, that basis allowance would, in all likelihood, not be -$2.00. It may actually be +$2.00 or +$1.00, reflecting the fact that a load of cattle averaging 700 lbs. is going to typically sell for a higher per-hundredweight price than a heavier load of cattle averaging 800 lbs. *The basis adjustment needs to be for the projected end weight of the cattle and for, as discussed earlier, the appropriate quality of the cattle as indicated by their grade.* Given the set of numbers that we have, there is a profit of $4.25 per hundredweight or, if you multiply that by 8 hundredweights, you are talking about $34 per head after all costs associated with this program have been covered. This $4.25 per hundredweight, or $34 per head, would be what the financial manager would think about as a contribution to investment, overhead, risk, and management (if a manager's salary has not already been prorated across the cattle and included in the variable costs). We assume in the table that the October feeder cattle futures are, in fact, sold at $85 per hundredweight on March 17.

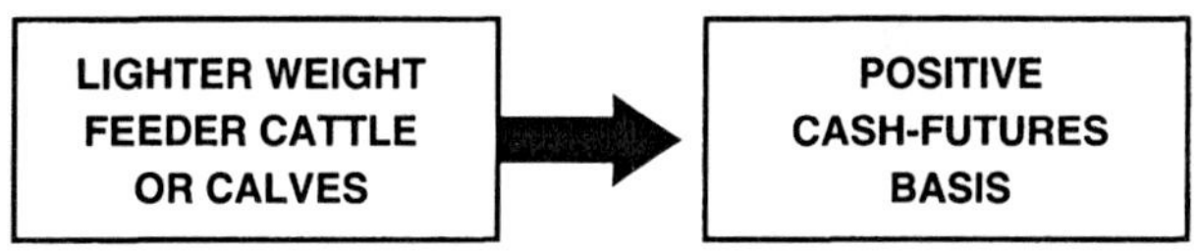

Lighter calves or feeders will be priced above 700-800 lb. steers

The reader should pay close attention to the September 25 example closing shown in Table 3. We are assuming here that the summer has been dryer than expected, cattle have not gained quite as well as projected, and there is also a need to get them sold in late September because the pastures are drying up. Thus the cattle weigh only 760 lbs. instead of the projected 800 lbs., but that is reflected in a somewhat better cash price relative to the futures than expected. Remember, we expected the cash cattle weighing 800 lbs. to sell $2.00 *under* the futures, and if they are lighter than that, say at an average weight of 760 lbs., they might sell somewhat more favorable relative to futures. Therefore, we are assuming that the closing basis actually turns out to be -$1.00, not the projected -$2.00. This means immediately that the overall net profit, instead of being $4.25 per hundredweight, will be $5.25 per hundredweight (assuming there are no changes in costs due to poor pasture conditions). It also means immediately that the realized price, which is made up of the cash price plus the net from the round-turn in the futures market, will be $1.00 higher than expected, or $84. *Go back through this to make* *sure you know where these "nets" come from--especially the +$8.00 in the futures market.*

Calculate the "nets"

These numbers are all shown at the bottom of Table 3. They should serve to remind us again that in the case of cattle, where gains cannot be predicted precisely and in the presence of a relatively variable basis (irrespective of gains and performance in the production program), the final results can be different than expected. Before we leave this point, it is important to recognize that if something had caused the basis to vary in a way that was *not* favorable to the producer, for example a closing basis at -$4.00 instead of the expected -$2.00, then the overall net would drop from $4.25 per hundredweight to $2.25 per hundredweight. The realized price would then drop from the expected $83 to $81 per hundredweight.

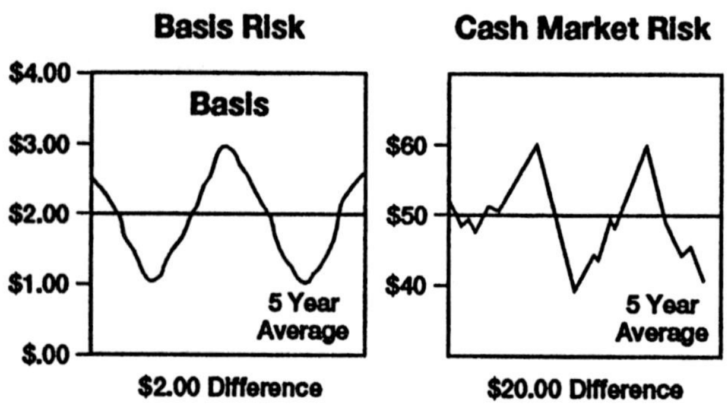

Even a variable basis means risk of basis variations much less than the risk of cash price variations

The reader is encouraged to think through this possible ending basis of -$4.00 and what it would mean before proceeding. As you gain experience, you will come to recognize that when the final basis is not the expected basis, you can adjust the overall net profit and adjust the realized price immediately without having to go back and do the arithmetic step by step. It might be useful to reiterate that if an $83 forward price was there when a basis allowance of -$2.00 is used, then an actual or realized basis at the close of the program of -$4.00 is going to mean a realized price of $81. An actual or realized closing basis at the end of the program of $0 would mean a realized final price of $85 per hundredweight, $2.00 better than the expected $83. The $2.00 "better" is because the basis is $0, not the expected -$2.00.

One of the conclusions that we might draw here is that if you go through the arithmetic and find that the forward price (and the profit) is small and not significantly different from breakeven, you then need to recognize that the lack of reliability in the closing basis could put the program into a loss situation. Many producers translate this into the idea that unless they can hedge a significant profit, they either don't undertake the program or they wait during the growing season and try to catch a better pricing opportunity later. Obviously, this latter approach means that they are starting through the year as a cash market speculator and are totally exposed to all of the losses that can occur if the market goes substantially lower.

The ending program shown in Table 3 reflects a lower cash market and the profits came from the futures account. If the cash market had gone to $86 and the closing basis was -$1.00, which means the futures had to be bought back at $87, the overall net would still be $5.25 per hundredweight, and the realized price would still be $84. But this net profit and the realized price would come from somewhat different sources. Selling cash cattle at $86 means a profit of $7.25 in the cash market. Buying back futures that were sold at $85 at $87 means a $2.00 loss in the futures account. The overall net when combined, then, is $5.25 per hundredweight. The realized price is made up of an $86 cash price minus a $2.00 loss in the futures, or $84.

Margin calls

The margin situation would be different in the latter case, however. When the markets go above the level at which the futures are sold, there will be a margin call once the maintenance margin threshold is triggered. The reader is encouraged to go back and work through the mechanics of Table 4 in Chapter 5 at this point, but it is also useful to discuss margins for feeder cattle. Since the contract calls for 50,000 lbs. live weight, each $1.00 move against you in the futures account amounts to $500. In the example discussed above, the move from $85 to $87 would mean an account loss in the futures of $2.00 per hundredweight or $1,000 per futures contract, and margin calls would have been issued as the market moved from $85 up toward $87. An initial margin requirement of $1,000 to $1,200 per contract with maintenance requirements of $600 to $750 would be fairly typical.

Consider feeder cattle margin calls

What this means is that the three-way agreement between the producer, the broker, and the banker, with the banker agreeing to provide an additional credit line for margin calls, is very important. *If that additional credit line for margins is not available for whatever reason, then the producer should think twice before using the futures to forward price the feeder cattle.* In feeder cattle, the availability of cash contracts is much less than was the case in grains. However, there are increasing possibilities that some broker, order buyer, or final user of the feeder cattle, such as the feedlot, would be willing to offer you a cash contract for delivery of feeder cattle of some specified weight range and specified quality. Those buyers, of course, would then typically take a position in the futures market as needed to protect their risk exposure. But it may be more difficult to find cash contracts for feeder cattle than it will be for the grains or for slaughter cattle or slaughter hogs.

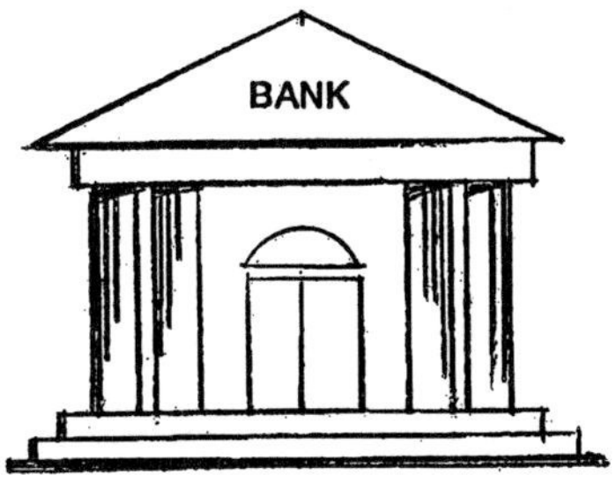

You must have a line of credit with your bank

ALTERNATIVES TO HEDGING IN THE FUTURES

Any alternatives to the futures in the feeder cattle market are therefore more likely to take the form of *formula pricing contracts* or basis contracts. A feedlot, for example, might be willing to forward buy your load of feeder cattle at a price that is $2.00 below the quoted price in some competitive auction market such as the Oklahoma City feeder cattle auction. Price quotes for 600 to 700 and 700 to 800-lb. cattle out of the Oklahoma City auctions are widely available and widely watched. The contract might call for a certain price relative to the Oklahoma City price, say "$2.00 under" for cattle that weigh on average 750 lbs. What the cattle buyer calls a "slide" might then be included in the contract to show how price will go up if the average weight is, say, 10 lbs. or more below 750, and go up even further if the average weights are 20-30 lbs. below 750. Conversely, prices would go down if the average weights are in the 760-770 range, and go down even further if they are 770-780, etc. This type of contract offers some assurance to the producers that they will get a price that is $2.00 under a widely watched and competitive Oklahoma City market. *It is important that the producer recognize, however, that he still has no protection against declining prices. What if the Oklahoma City prices plunge?*

PRICE RISK

To go back to our example, we needed a price for these steers in the early fall months of $78.75 for the program to break even. In March, the 700 to 800-lb. feeder cattle in Oklahoma City may be quoted at $88 per hundredweight, and the producer thinks that if he has to take $2.00 less or $86, that still looks real good. But this is a March price. If the weather turns dry and corn costs to the feedlot go higher during the summer months, they will try to buy the feeder cattle lower. There is absolutely nothing therefore, to keep the Oklahoma City market from going down to, say, $75 and the producer's effective price from dropping to $73. Major losses have been incurred. *Pegging the price relative to some observable market via a formula pricing arrangement is not the same thing as getting protection against declining prices.* It does peg the relationship to an observable market, but the producer is still exposed to price risk.

I must remember I still face pice risk

BASIS CONTRACT

The other type of contract that might be available is, as suggested above, a basis contract. A buyer of feeder cattle might be willing to pay, for example, $1.00 under the October feeder cattle futures. This specifies the cash-futures basis in the final pricing arrangement. That is why it is called a *basis contract*. Again, however, the producer is not protected against a declining market. If the March trading level of the October futures is $88 and he can negotiate a price no more than $1.00 under $88, that $87 price looks good relative to the break-even in the table. *But there is nothing to keep the October feeder cattle futures from going to the mid-to-low $70s, and if that happens, a loss will be incurred.* There is often some additional flexibility in this type of contract, however, and many producers are interested.

A basis contract secures a good basis, but you are still facing price risk

The flexibility comes in the form of the buyer's typical willingness to allow the seller to pick the day to price prior to, typically, October 1, the beginning of the month for which the futures are being traded. This puts the producer in a position of monitoring the market and watching it trade from perhaps the high $80s, and down to the low $80s, back to the middle $80s, depending on what the fed cattle market is doing and what is happening to corn costs. It is very important, then, that the producer be in a position to intelligently monitor the markets and make an objective decision about when the best price that is going to be offered is probably being offered and to pull the trigger and set the price. The buyers take whatever position they need to take in the futures market to protect themselves, and from that point forward, the seller does in fact have price protection.

For example, on a day when the October feeder cattle futures are trading at $86, if the seller calls and says he wants to establish a price that day and effectively establishes a price at $85 (remember, $1.00 below the futures), then a specific price has been set. The buyer will need to go short or sell the underlying October feeder cattle futures to protect against paying $85 for cattle that, when delivered in late September or early October, might be worth only $75 per hundredweight. But this is no longer the seller's problem. The *buyer* is having to manage the positions in the futures complex and assume any risk the cash and futures markets will not show a basis of-$1.00. The seller has not only now pegged a specific price level at $85, he is no longer subject to any basis risk since the effective basis of -$1.00, that he was being quoted in the contract, has been captured via the $85 price.

SPECIFICS ON FEEDER CATTLE FUTURES HEDGING

Before proceeding to some illustrations of hedging slaughter cattle and hogs, it is useful to reflect on the unique characteristics of hedges in feeder cattle. What makes this particular situation different includes the following:

- **The basis for feeder cattle is typically more variable and less predictable than the basis for slaughter cattle or slaughter hogs. When the cattle do not weigh the 700 to 800 lbs. for which the feeder cattle futures are designed, the variability in basis can be still greater.**
- **Performance during the production period is probably more uncertain in feeder cattle than for slaughter cattle or slaughter hogs since weather can influence quality of pastures and availability of forages.**
- **There are a significant number of weight differences and quality differences that can and will influence the closing basis feeder cattle.**
- **The non-futures contract market, which usually includes cash contracts and basis contracts, is probably less widely available for feeder cattle than for slaughter cattle, slaughter hogs, or for the grains and oilseeds.**
- **There will be far more problems meeting contract size requirements in feeder cattle than in slaughter cattle or slaughter hogs.**

HEDGES FOR SLAUGHTER CATTLE

Procedure for forward pricing or hedging slaughter cattle is much the same as that for feeder cattle. Nonetheless, it is useful to present an example just so the differences in cost of the final product become clear and to reinforce the reader's understanding of basic procedure. Table 4 provides an example using a format very similar to that employed in earlier examples. There is a date column, a column for what is going on in the cash market, and a column for what is going on in the futures market.

THE DETAILS

The top of Table 4 provides some detail on what is involved. The 750-lb. steers can be bought at an in-lot cost of $84 per hundredweight or $630 per head. This is just the purchase cost of the steers, plus trucking into the feedlot, an allowance for shrink during transit, etc. Some allowance for death loss and weight loss due to sickness on some percentage of the cattle would also need to be applied. This particular set of steers is projected to gain 3 lbs. per day at a feed cost of $.55 per lb. That adds another $198, given that they will be fed for 120 days. Adding 360 lbs. to the initial weight of 750 lbs., generates a 1,110 lb. steer. Note that we are assuming the typical practice in the southwestern part of the U.S., where the steers are sold with a 4 percent shrink. This means feedlots are actually paid for 1,110 lbs. less 4 percent of that level or 1,066 lbs, which is 10.66 hundredweights. This means if you add the $630 initial purchase cost, the $198 cost of gain, and divide by what would really be the weight for which you will be paid (10.66 hundredweights), the break-even is $77.67 per hundredweight. Based on past records for this particular type of cattle in your local market area, you expect the closing basis to be -$1.00. The forward price is then $81.00, and the profit possibility is $3.33 per hundredweight or $35.50 per head if you multiply $3.33 times the 10.66 hundredweights for which you will be paid.

This possibility looks quite attractive, so the cattle feeder decides to go ahead and buy the feeder cattle. He is still able to get them bought at $84 per hundredweight on August 11, and he finds that December futures, although they have varied some since August 10, have continued to trade around $82. He is able to get those December futures sold on August 11 at $82.00 per hundredweight to establish the price protection.

Date	In Cash	In Futures
August 10	Can buy 750-lb. steers at a "laid in" cost of $84 per cwt. or $630 per head. Project gains at 3 lbs. per day at a cost of $.55 per lb. or $198 for the 120-day feeding period. Break-even on 1,110 lbs. steers sold with a 4 percent shrink will be ($630 + $198) ÷10.66 cwt. or $77.67 per cwt. Estimate cash-futures basis for finish date near December 10 at -$1.00 per cwt. FP = $82.00 - 1.00 = $81.00 Profit = $81.00 - 77.67 = $3.33 per cwt. ($35.50 per head)	December live cattle futures trading at $82.00.
August 11	Buy feeder cattle @ cost of $84.00	Sell December futures at $82.00.
December 8	Sells steers weighing 1,110 lbs. on average with a 4 percent shrink at $79.00 per cwt. Net: $1.33 per cwt.	Buys back December cattle futures at $79.50. Net: $2.50 per cwt.
	Overall Net: = $1.33 + 2.50 = $3.88 per cwt. Realized Price = $79.00 + 2.50 = $81.50	

Costs of hedge: Commission costs of $.10-$.15 per cwt., plus a nominal interest cost on about $1,000 initial margin deposits.

Table 4 - Demonstration of a short hedge for slaughter cattle

Cost in a cattle feeding program

Obviously, the primary cost in this program is the feeder steer. The 750-lb. steer that cost $84.00 per hundredweight delivered costs $630--and is truly the big cost component. We will come back to this point later when we talk about using long hedges to try to keep the cost of these steers going into the feedlot under control. The cost of this steer during the 120-day stay in the feeding period should always include an interest cost on the $630 that is tied up in each steer. Typically, most decision makers would also apply some interest charge to the cost of gain including, of course, the cost of the feed. If that feed is being bought on a weekly basis, you might reasonably apply an interest cost to about half of the $198 cost of adding the gain. All of that we assume has been included in the $630 initial procurement cost and the $198 cost of holding that inventory and feeding it. It is all reflected, therefore, in the break-even of $77.67 per hundredweight.

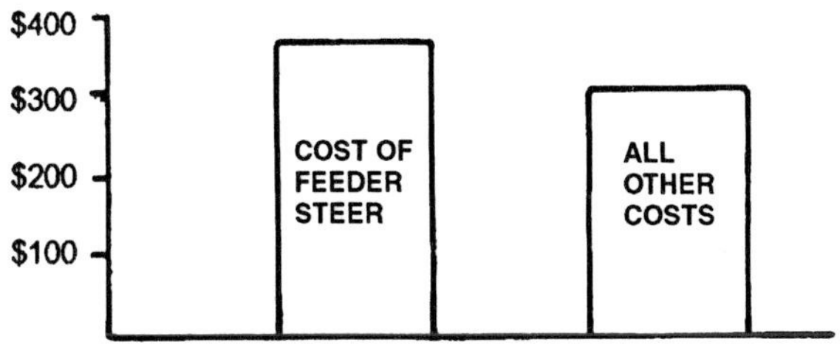

Graph 1 - The cost of the feeder steer is the big cost in cattle feeding programs

The bottom portion of Table 4 shows a possible ending if the cattle are sold on December 8. Say the cattle feeder is able to sell the steers that actually weigh the expected 1,110 lbs. with a 4 percent shrink at $79.00 per hundredweight. The pleasant surprise comes when the cattle feeder is able to call his broker and get the December cattle futures bought back at $79.50. This reflects an actual closing basis of -$.50, not the -$1.00 that has been allowed for. We know immediately that our nets are going to be higher, and we find that when we add the $1.33 from the cash side to the $2.50 from the round-turn trade in the futures, we have $3.88 per hundredweight. That is $.50 per hundredweight more profit than we had expected, and the realized price is $81.50, $.50 more than we had projected the realized price would be. All of this occurs, of course, because the basis variation in this particular instance was in favor of the producer--closing at -$.50 instead of the expected -$1.00.

Basis and basis risk

The reader should back away from this for just a moment and reflect on what would happen if the cash cattle were sold at $79.00 and the best the producer could do in buying back the December futures was $80.50. Then, you should note the basis variation has worked *against* the producer. We will find that instead of the expected price of $81.00, the actual realized price will only be $80.50 and that the actual profits will be down by $.50 per hundredweight compared to the $3.33 that had been factored into the decision. All of this happens, of course, because the basis did not move to exactly the level that had been allowed for in the example. *The basis levels in the fed cattle market are more predictable than those for feeder cattle, but they are still variable.*

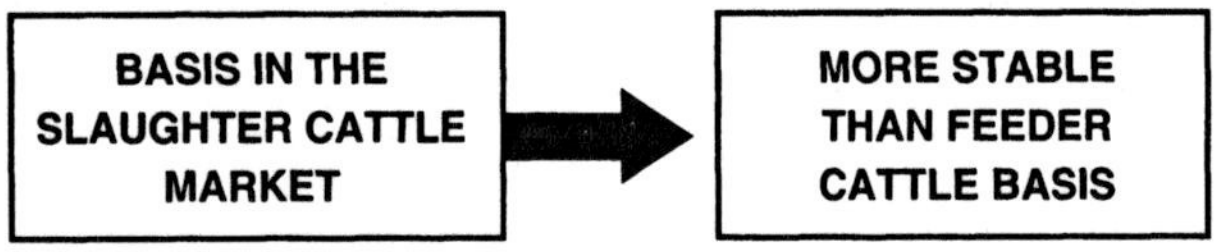

Basis performance in live cattle futures is usually good

Also, the bottom of the table shows some simplistic estimates of the cost of the hedged program. Commission costs can run anywhere from $35 to $80-90 per round-turn on the 40,000-lb. contract, so a reasonable estimate of those costs is $.10-.15 per hundredweight. There should probably also be at least a nominal charge assessed against the $1,000 margin per contract (or something close to this level) that you have tied up with the broker in the form of margin deposits. Keep in mind that if this market had traded up substantially and gone to, say, $85-88, the cost of the hedge would be higher because there would be interest on more money that was tied up in answering the margin calls. If the basis turns out to be at the end of the program -$1.00, the net would still, of course, be the same at $3.33 per hundredweight as had been expected, and the realized price would still be the expected $81.00 per hundredweight. The money that goes to the broker in the form of margin calls does not come back to the producer, it should be noted. It is "lost" and is covered by the unexpectedly high value of the cash price. *The only difference in the final position, when the market goes up against the short hedge position, is that the interest cost on margin monies will be slightly higher.*

The interest on margin calls is not worrisome

HEDGES FOR SLAUGHTER HOGS

There are two types of slaughter hog producers. Historically, it was common for a producer buy 40 to 60-lb. feeder pigs, place them on a feeding floor, and feed the hogs to slaughter weights. Procedure for a short hedge in this instance is exactly the same as that just discussed in Table 4 for slaughter cattle. In other words, a feeder pig is bought and then fed to slaughter weight, so the cost of producing that slaughter animal is a function of what you have to pay for the feeder pig plus your costs of the gain.

FARROW-TO-FINISH

The hog industry continues to shift toward a farrow-to-finish operation as compared to an operator who finishes slaughter hogs as a feeding floor specialist. The costs, then, are not a function of market prices for feeder pigs but rather a function of costs of producing a slaughter hog from an operation that covers all the functions from the time the pig is born until the hog is sold at slaughter weight. This type of operation has already reached strong majority status in the industry and is expected to trend even higher. Thus, in Table 5, the demonstration of a short hedge for slaughter hogs treats the situation that would face a farrow-to-finish producer.

This particular operator projects the costs of producing a 240-lb. slaughter hog to be sold in early October at $41.50. This is a break-even price that ought to be reached by a producer of even moderate size who is an efficient operator so long as feed costs are reflecting a $2.25 to $2.50 per bushel corn market. Based on past records in this particular market area, our producer expects the basis in early October to be around -$.75 per hundredweight.

Reports that change supply expectations can bring drastic price changes in hogs

The producer observes in late June that October live hog futures are trading at $48.25. He has seen the October futures trade up with a strong seasonal surge in cash hog prices and corollary increases in the prices for the July and August futures. This surge has carried the summer market into the mid-$50's. Drawing on his understanding of the seasonal pattern in hog prices and expecting the best forward pricing opportunities for October to occur in the early to mid-summer months, the producer decides to take advantage of this profitable opportunity. He is concerned that the upcoming Hogs and Pigs report will show expansion throughout the industry, significant increases in slaughter in the fall months,

Date	In Cash	In Futures
June 20	Project costs of producing 240-lb. hogs to be sold in early October at $41.50 per cwt. Expected closing basis in early October at -$.75 per cwt. FP = $48.25 - .75 = $47.50 Profit = $47.50 - 41.50 = $6.00 per cwt.	October live hog futures trading at $48.25
June 20		Sell October hog futures at $48.25
October 3	Sell cash hogs at $39.00 Net: -$2.50 Overall Net: = -$2.50 + 8.50 = $6.00 per cwt. Realized Price = $39.00 + 8.50 = $47.50	Buy back October hog futures at $39.75 Net: $8.50 per cwt.

Table 5 - Demonstration of a short hedge for slaughter hogs

and the possibility of prices at or below his break-even levels. Thus, he sells an appropriate number of contracts to cover the 50,000-lb. truckload lots that he expects to sell in early October.

Contract size

Our producer is relatively new at hedging, and he discovers that each live hog futures contract is for 40,000 lbs. or 400 hundredweight. He is a relatively small producer with about 200 sows, but he can put together a truckload lot of hogs every few weeks. He has one of these scheduled for completion in early October. What this producer might do, then, is sell one of the live hog futures contracts recognizing that this only covers 40,000 lbs. of his 50,000-lb. load. (The reader should realize that as producers get larger and have more volume to sell, then some multiple of the 40,000-lb. contracts traded on the Chicago Mercantile Exchange would need to be involved to forward price all the hogs.)

A truckload of hogs is 50,000 pounds, which is more than the 40,000-pound hog futures contract

Returning to our example, we find that on October 3 when the hogs are sold, the producer's concerns of late June have been justified. The cash market has been pushed down to $39 per hundredweight because of a 6-7 percent year-to-year increase in hog slaughter and almost 10 percent increase in poultry production coming to the market in the fall months. Note that the cash market speculator who did no forward pricing and had the same costs would, in this instance, lose $2.50 per hundredweight. That is about $6.00-6.50 (2.40 cwt. times $41.50 minus $39.00) per hog if he had to sell in a $39 cash market. Remember, the individual producer is a price taker and would not be able to negotiate a price significantly different from the $39 level.

Our producer has a hedge, however, and when the cash market goes down, the futures market goes down with it. The October 3 ending illustration assumes that the basis goes exactly to the -$.75 per hundredweight that had been expected. Under those conditions, before you do the arithmetic, you know that the profit is going to be $6.00 per hundredweight before the cost of the hedge is deducted and that the realized price is going to be $47.50. The arithmetic is, in fact, shown in the bottom of the table. Note that the overall net comes from the -$2.50 in the cash market and the +$8.50 from the round-turn in the futures market, giving a profit of $6.00 per hundredweight. The realized price, as always, is the cash price at which you have to sell plus the net from the round-turn in the futures, or $47.50. Remember, there would be $.10-.15 per hundredweight costs of the hedge, and that would pull the net profit per hundredweight down more nearly to the $5.85-5.90 range as compared to the $6.00 before hedging costs are deducted.

REALIZED PRICE = **CASH PRICE** + **NET FROM FUTURES TRADES**

The realized price from a hedge is cash plus the results from the round-turn in futures

KNOWLEDGE AND OPPORTUNITY

It is useful to step back and recall the discussion about price movements over time in Chapter 2 and the micro-macro trap in production agriculture mentioned in that Chapter and discussed in more detail in Chapter 3. *Here, we have an excellent example of the importance of understanding those time-related patterns in price. Our producer took advantage of the seasonal surge in the hog market to seize the opportunity and sell October futures at a favorable level.* Notice that the $48.25 level at which he was able to sell the futures was still several dollars per hundredweight above the summer market that had reached $55-56 per hundredweight. This is normal, and we have to keep in mind that the price being discovered for October delivery at the Chicago Mercantile Exchange is being discovered by traders who understand the seasonal patterns in price. It is just that when the market goes up and is pushed higher by reduced slaughter levels in the summer months, the more distant futures often will move up at least a modest amount with the seasonal strength in the summer hog market. Thus, an opportunity is created, and the producer is in a position to take advantage of his knowledge of the seasonal pattern in prices.

MANAGING THE PRICE CYCLE

Also discussed in Chapter 2 and again in Chapter 3 was the really troublesome price cycle in hogs. We see an additional example of it here with the implication that not only was there some decline going into October in a seasonal basis, but much of that price decline came because there was a cyclical expansion going on in hogs that was not known until a quarterly report was released near the end of June. This ties back to the discussion in Chapter 2 where a producer is well advised to look ahead and forward price hogs as far into the future as possible.

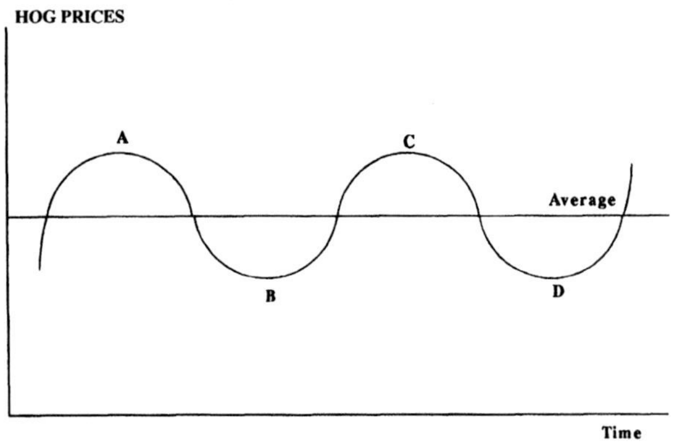

Graph 2 - Try to forward price at points A and C to protect against the price cycle

The key is to anticipate at least within a few months' time when prices will turn down under the weight of a cyclical expansion. The reader is reminded that while Table 5 illustrates with one load of hogs to be sold in October, this same small producer could have forward priced hogs to be sold in November with the December futures, or hogs to be sold in late December and January with the February futures from the next year, etc. *On June 20, he would have been able to to forward price those hogs and get price pegged at profitable levels before the cyclical expansion pushed prices down and wreaked havoc on all those producers operating as cash market speculators.*

LONG HEDGES ON STOCKER AND FEEDER CATTLE

Long hedges on light cattle to move into either pasture or feedlot programs can be very important. It is an opportunity again to take advantage of understanding of the seasonal pattern in prices and to get away from the micro-macro traps that often loom for the unsuspecting producer. In this section, we will look at both a long hedge on light calves moving into a backgrounding or stocker cattle program, as discussed earlier in the chapter, and at 700 to 800-lb. yearling steers or heifers moving into a feedlot program. The opportunity to do something about cost of cattle at the onset of these programs is important to both types of buyers.

Table 3 showed a 450-lb. steer calf coming into the stocker program at a laid-in cost of $100 per hundredweight. This means $450 is tied up immediately in the cost of the calf. This also means that during periods in which the purchase price of those calves is relatively high, the producer whose business it is to turn 450-lb. steer calves into 800-lb. yearlings faces a tremendous amount of risk. This is especially true during years in which weather or other problems threaten the corn crop. Thus, to all of the other sources of price uncertainty facing the feeder cattle producer, we need to add the weather uncertainty that can have an impact on corn costs.

Before proceeding, let's pursue that latter point in more detail. Remember, a feedlot operator faces two big costs in producing slaughter cattle. The first, and typically the biggest, is the cost of the feeder steer or heifer. The second, and very important, is the cost of the grains to feed that animal during the 120 to 150-day stay in the feedlot. As a rough rule of thumb, it will take about 50 bushels of corn to carry a 750-lb. steer or heifer up to a 1,100-lb. selling weight. This will vary from feedlot to feedlot, but for illustrative purposes, let's assume it takes 50 bushels of corn.

FEEDER CATTLE VERSUS CORN COSTS

If the market price of that corn goes up $.10 per bushel, then clearly the feed costs, considering corn alone, go up $5.00 per head immediately (50 x $.10 = $5.00). But what if the corn market, because the weather threatens the crop substantially, goes up $1.00 per bushel? Then, instead of $5.00 additional cost on feeding that steer, we are talking about $50 per head (50 bu. x $1.00) in additional corn costs alone. Typically, the cost of protein supplement such as soybean meal, the cost of hay, and the cost of any silage that is used in the ration formulation would also be going up as corn costs go up. So, a $1.00 increase in corn costs could add $60-75 per head to the total feed cost of producing a slaughter animal from a yearling steer or heifer.

HIGHER CORN COSTS → **LOWER PRICES FOR FEEDER CATTLE**

If corn prices increase, feedlots have fewer dollars to bid for feeder cattle

Let's assume that the increase in feeding costs is, in fact, $75 per head. What does that mean in terms of what the feedlot operator will be able to pay for a 750-lb. steer? If he has incurred a $75 increase in feeding costs, other things being equal, he has to try to buy the 750-lb. steer $75 cheaper. Since we are dealing with 7.5 hundredweight here, it is not exactly complicated math to recognize that the buyers are likely to push the price of the 750-lb. feeder steer down by around $10 per hundredweight. If the producer who is running the pasturing program or a roughage-based backgrounding program owns the 450-lb. calf and is growing him up toward 750 lbs. while corn goes up $1.00 per bushel, that producer can face major losses.

BE AGGRESSIVE

The break-even cost on the finished 750 to 800-lb. steer may be in the low to mid-$80's if he paid $1.00 per lb. or $100 per hundredweight for the 450-lb. calf. If corn goes up sharply, feeder cattle prices are likely to be driven to the low $70's, possibly the $60's, and major losses are pending. *Thus, it is very important that the producer of feeder cattle who holds cattle on grass or in backgrounding programs during the summer months be aggressive in getting those cattle forward priced when profitable opportunities develop.* That means, of course, that the forward pricing illustration laid out in Table 3 can be especially important, and the astute producer will be ready, willing, and able to lock in a decent profit on his feeder cattle if and when the market offers it.

WEATHER UNCERTAINTY IN CORN CROP → **RISK FOR OWNER OF CATTLE**

Weather risk in corn production means price risk on feeder cattle during summer months

Returning to our example of a long hedge, Table 6 lays out a sample long hedge when the need of the producer is to buy 450-lb. steer calves in late February to carry through the spring and summer months and into early October on a pasture program. Note the procedure here is much like the long hedge on corn that was discussed in Chapter 6. In this case, however, the reader would be well advised to pay particular attention to the basis employed and to remember that basis variability may be more of a problem in feeder cattle than in most other commodities with which we deal. In fact, the ending of the program laid out in Table 6 demonstrates what can happen when the basis does not turn out to be what you expect. In that particular instance, the cost of your 450-lb.

Date	In Cash	In Futures
October 15	Will need to buy 450-lb. steer calves in late February. Estimate the cash-futures basis for late February at +$15.00 per cwt. Forward Cost = Futures + Basis FC = FUT + BASIS = $77.00 + 15.00 = $92.00 per cwt.	March feeder cattle futures at $77.00
October 15		Buy March feeder cattle futures at $77.00
February 25	Buy 450-lb. steer calves at $100.00 per cwt. Net Cost = Cash - Futures Net = $100 - 7.00 = $93.00 per cwt.	Sell March feeder cattle futures at $84.00

Table 6 - Demonstration of a long hedge on 450-Lb. steer calves in a pasture program steer calves turns out to be higher than you had anticipated because the basis was wider than you had expected it to be when the hedge was placed.

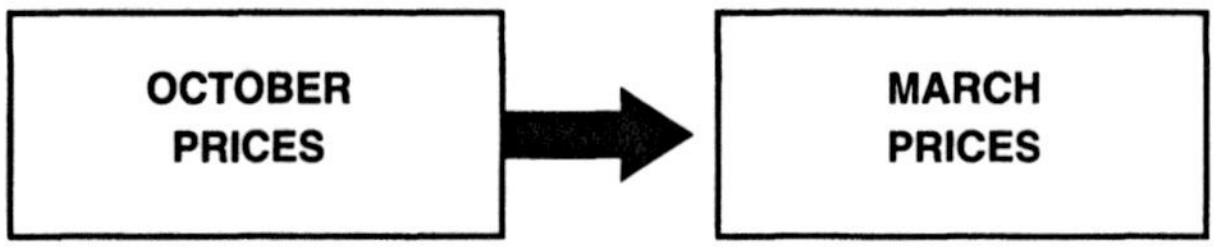

Knowing the seasonal price pattern from October to March can guide long hedge decisions

BASIS LEVELS

The reader should not be bothered that the basis appears to be a very big $15 per hundredweight. So long as corn is at or near a normal price of $2.50-2.75 per bushel, the light calves will always sell at a substantially higher price per hundredweight than, say, the 700 to 800-lb. yearling steers. The steer calves are in demand by stocker operators, which is what we are demonstrating here, and are also in demand to go directly into feedlots. What we are seeing, in the top part of Table 6, are cash prices of 450-lb. steer calves projected at around $15 per hundredweight above the trading level of the March feeder cattle futures in February. Be reminded that the feeder cattle futures are trading yearling steer cattle that weight 700-800 lbs. We are dealing with a different size and weight of animal here. That is another reason these cash-futures basis relationships are likely to be somewhat variable and less than perfectly predictable.

At any rate, the March feeder cattle futures are trading at $77 on October 15, and our producer expects to see those feeder cattle futures trade up. Once again, we find someone taking advantage of their knowledge of seasonal price patterns. The producer knows that cash prices are typically lowest in the fall, when cattle come to market off grass, and will tend to trade up moving toward the following spring. The Chicago Mercantile Exchange trader knows this too. The 700 to 800-lb. yearling steers might be at $74 on October 15, and March feeder cattle futures are reflecting a seasonal price pattern and trading at $77. What our producer expects, however, is that the entire price structure will move up in the winter months with no uncertainty left from the corn side of the equation and the possibility of decent fed cattle prices appearing likely. If we combine the $77 trading level for the March feeder cattle futures and the $15 basis, then the forward cost for the 450-lb. steer calves is $92 per hundredweight. This is fairly routine procedure by now and involves combining the futures and the basis to get, in this instance, a forward cost instead of a forward price as has been demonstrated in many of the earlier examples where a short hedge was in order.

We assume the producer buys March feeder cattle futures at $77 on October 15, and then we move to a possible ending scenario on February 25. Everybody else is trying to buy these light calves to go on grass in late February, and the producer can purchase the quality and type of calf that he wants no lower than $100 per hundredweight. On that same day, he calls his broker, finds that the March feeder cattle futures have been trading a few cents above $84, and down to as low as $83.10. He makes the decision that a sale order at $84 might be filled during the trading period for the contract, and does, in fact, place such a sale order with the broker. Later in the day, the broker calls the producer and informs him that the order has been filled and the March feeder cattle futures were sold at $84. When we combine the net cost of $100 in the cash market and a round-turn net of $7.00 per hundredweight in the futures market, the net cost of the cattle is $93 per hundredweight.

GETTING TO NET COSTS

Two additional points need to be made here. The first is that the reason the $93 is present instead of the expected $92 is because the closing basis is $16 per hundredweight and not the expected $15 per hundredweight. The second point is that the $7.00 net from the futures round-turn is subtracted from the cash cost, not added to it. *This is because we are talking about buying an input, and any benefits that we generate via the round-turn in the futures market will be used to reduce the costs that have to be incurred in the cash market.* Thus, we get to $93 per hundredweight. The producer is substantially better off than the cash market speculator who has to pay a net cost of $100 for the same type of steer calves.

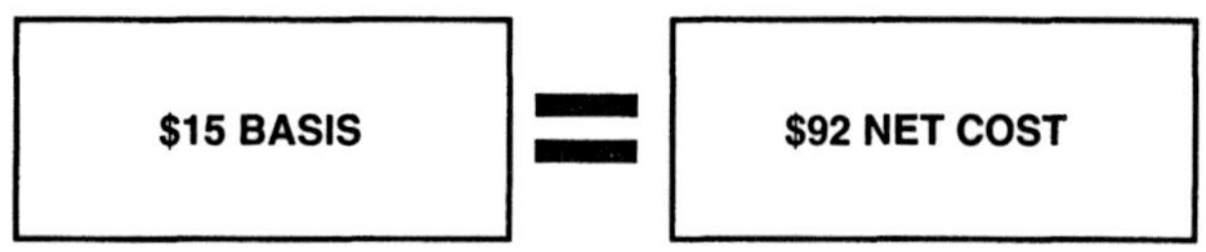

The large positive basis for 450-lb. steer calves is normal, and the long hedge works

Table 7 demonstrates the position of a cattle feeder whose program involves feeding heifers to slaughter weights. It is set up with the same basic price levels in futures and the same dates as the example in Table 6 just to remind the reader that there are various types of decision makers that use these markets. In this particular instance, the operator of a feedlot is going to have to buy heifers weighing somewhere near 750 lbs. in late February. They will replace a pen of cattle that he projects will be ready for sale (because they are at slaughter weights) in late February. He can take advantage of his understanding of seasonal price patterns, and he thinks that the March feeder cattle futures trading at $77 are likely to trade higher as we move toward March. By the time he applies a -$8.00 per hundredweight basis expectation, he realizes that he can peg his cost of heifers to be bought in late February at $69 per hundredweight.

Date	In Cash	In Futures
October 15	Will need to buy 750-lb. heifers in late February to replace a pen of cattle projected for sale in late February. Estimates late February cash-futures basis at -\$8.00 per cwt. FC = FUT + BASIS = \$77.00 - 8.00 = \$69.00 per cwt.	March feeder cattle futures at \$77.00
October 15		Buy March feeder cattle futures at \$77.00
February 25	Buys 750-lb. heifers at \$76.00 Net Cost = Cash - Futures Net = \$76.00 - 7.00 = \$69.00 per cwt.	Sell March feeder cattle futures at \$84.00

Table 7 - Demonstration of a long hedge for 750-Lb. yearling heifers in a feedlot program

A HARD LOOK AT BASIS

Before proceeding with the demonstration in Table 7 let's back away and discuss the -\$8.00 basis. For any number of reasons, steers of the same weight and grade tend to sell at a higher price than heifers when we are talking about feeder cattle. Heifers do not gain quite as well in the feedlot as steers and are generally reputed to produce carcasses with slightly inferior yield grade or cutability characteristics as steers from the same cattle herd. What we are seeing here suggests that the cash market for heifers in late February has, based on past records, tended to run \$8.00 per hundredweight below the trading level of the March feeder cattle futures.

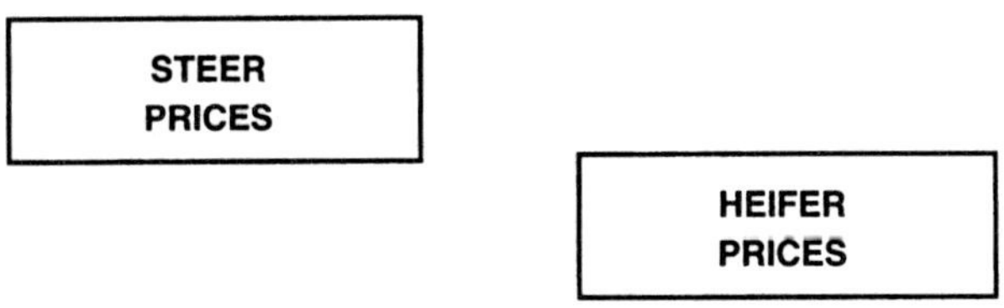

Heifer prices will be below steer prices when the cattle weigh the same

Given this, the forward cost, then, is \$77 minus the \$8.00 basis (actually, you add the negative \$8.00) or \$69 per hundredweight. The producer likes the looks of this, and he buys March feeder cattle futures at \$77.

The closing date is, again, February 25. As was the case in Table 6, the March feeder cattle futures are trading around \$84. In this particular instance, our feedlot operator goes into the cash market and is able to buy the 750-lb. heifers that he wants at \$76. He calls up his broker and sells March feeder cattle futures at \$84. The net cost of the cattle, then, is what he has to pay in the cash market minus, the net from the futures round-turn--\$76, minus \$7.00 per hundredweight or a final cost, as had been anticipated, of \$69 per hundredweight.

It should be apparent to the reader that if the market goes down such that the March feeder cattle futures have to be sold below the \$77 purchase price, then the net turn in the futures market round-turn trade is negative. *Under these circumstances, that negative round-turn from futures has to be added to the cash cost of the cattle to get the final cost.* Again, if the basis is the expected -\$8.00 per hundredweight at the close of the program, the final cost will be \$69. Readers might want to work through an example here to make sure they understand where the various components of the final cost would come from if the cash market goes, for example, to \$66 and the futures are sold at \$74 instead of \$84. This way, the net cost of the cattle is the \$66 that has to be paid in the cash market to get the cattle, plus the \$3.00 loss that is incurred in the round-turn trade in the futures.

CASH PRICE PAID FOR CATTLE + **LOSSES FROM FUTURES TRADES** = **COST OF THE CATTLE**

When prices move down, losses from long hedge positions must be added to the cash market price

As is always the case, there are some costs associated with the hedge. Commission costs are roughly $.10-.15 per hundredweight, as they were with the live cattle futures. The weight of the modern feeder cattle futures contract is 50,000 lbs., essentially matching a semi-truckload lot. The user of the market needs to consider the fact that each contract is for 50,000 lbs., or 500 hundredweight, in deciding how many contracts he needs to cover either his steer calf needs, his yearling steer needs, or his yearling heifer needs, as demonstrated in Table 7.

Options are flexible, but a premium has to be paid as added costs

PRICING WITH OPTIONS

Obviously, feeder cattle, slaughter cattle, or slaughter hogs could be forward priced using options. This establishes a price floor and leaves open the possibility of higher prices in the event they should occur. The procedure for the options was illustrated earlier in Chapters 3 and 4, but it is worthwhile to stress at this point the important difference between pricing directly in the futures market and buying a put option as a producer. Examples using the option approach will be included in the case studies at the end of the chapter.

PRICE FLOOR

If we assume for the moment that the strike price picked by the user is exactly the same as the trading level for the underlying futures contract, *then the price floor set by using a put option is below the specific price you could establish in the futures by the amount of the option premium.* In other words, the forward price for slaughter hogs if the futures contract is trading at $50 is established as follows:

Forward Price = Futures + Basis

FP = FUT + BASIS.

If the basis is -$1.00, then the forward price is, of course, $50 minus $1.00 or $49 per hundredweight.

If the put options are used and a $50 strike price is employed, then by way of review we have the following:

Floor Price = Futures + Basis - Premium

Fl P = SP + BASIS - PREM.

What we see here is that the strike price replaces the futures level used in the first example, and the additional component in calculating the floor price is the option premium. If we were dealing with a situation, for example, where the option premium was $1.50 per hundredweight, then obviously the floor price would be $50 less the $1.00 basis less the $1.50 option premium or $47.50. What this shows very clearly is that the floor price that can be established using a put option is below the specific price that can be established using the futures by the amount of the option premium. Be careful--the difference will be exactly the premium cost *only* when the futures price in the hedge example and the strike price in the option example are the same.

STRIKE PRICE INTERVALS

To emphasize that last point, the reader should not be confused by the fact that often the underlying futures contract is trading at some specific level such as $50.87 and the producer has a choice of strike prices that are only in even dollar amounts. In this particular instance, the $50.87 translates into a forward price of $49.87. If the producer knowing the market is trading at $50.87 makes the very rational choice of $50 per hundredweight for a strike price, then the floor price is still $47.50. Obviously, the specific price established in the futures is above the floor price established for options *by more than the premium because the futures price is $50.87 instead of the $50 strike price employed in the option considerations.*

If we turn the coin over and look at the cost side of the equation, then a similar discussion can be completed. If you have to buy feeder cattle as an input, you can establish a specific cost by taking the trading level of the appropriate futures contract and adjusting it for basis. If, for example, you wanted to peg the cost of feeder cattle you are going to have to buy in late February and you project the cash-futures basis to be -$1.50 per hundredweight, then if the March feeder cattle futures are trading at $84.50, you are pegging your costs at $83. This is simply a matter of taking the trading level of the underlying futures and adjusting it for basis, which in this particular instance is -$1.50 per hundredweight. The reader should keep in mind that the basis on feeder cattle will tend to be positive when you are dealing with smaller types of cattle. You would then have to *add* the basis to get the cost.

When you think about using call options to protect against rising costs, then the comparisons are similar to the discussion for the put option. The difference here is that we have someone who is having to buy feeder cattle, corn, or whatever the case may be as an input, and he is worried about costs going up. He can peg a specific cost, as we have just suggested, on the feeder cattle by buying the appropriate futures contract. Buying March feeder cattle futures at $84.50 with a basis expectation of -$1.50 pegs the cost of cattle to be bought in late February at $83. The only thing that could change that, we should realize by now, is if the basis turns out to be something other than the -$1.50 that has been incorporated into the calculations.

PREMIUM COSTS ADD TO COSTS

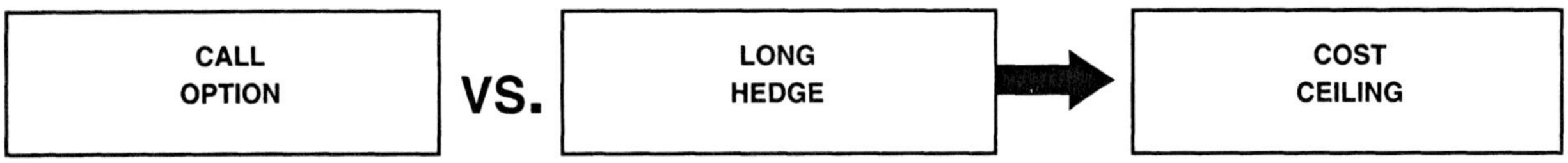

A call option is an alternative to the long hedge and sets a cost ceiling

If you use a call option instead of the futures directly, then you are imposing a *cost ceiling. It is very important to understand that in this instance the cost of the option premium has to be added to the cost of the cattle.* Let's assume, for example, that the March futures are trading at $83.50, and the cattle feeder decides to buy an $83 call option. This means costs of the cattle have been pegged at $83 less the $1.50 basis allowance plus the option premium. If we assume the option premium is $2.00 per hundredweight, then the cost ceiling imposed on the cattle in this particular instance is $83.50 ($83.00 - 1.50 + 2.00). Notice that this is somewhat higher than a specific cost that could be pegged by going directly into the futures, and it all depends on what strike price is employed relative to the trading level of the underlying futures market. Strike prices for feeder cattle are offered in even dollar increments, and you cannot, even if you wanted to, buy an $83.50 call option.

FUTURES VERSUS OPTIONS

At several points in the book, there is discussion about whether a decision maker should use the futures directly or use options. Some of this discussion will be incorporated into the case studies at the end of this chapter. *In general, you have to keep in mind that when you are using an option, you are paying for flexibility.* The producer who is worried about prices falling and buys a put option is paying for someone else to accept the risk that the market might, in fact, go up instead of go down. The user of corn or the buyer of feeder cattle who uses a call option is paying a premium to leave open to them the possibility that costs will actually come down relative to expectations. To some extent it is oversimplifying, but which approach should be used often depends on whether or not it makes any sense to be buying flexibility.

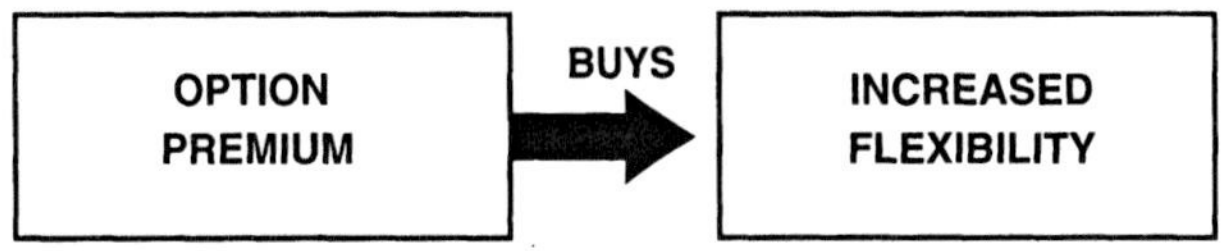

Option premiums buy you the opportunity to benifit if prices or costs move in your favor

For example, if you are talking about buying an $83 call on a feeder cattle contract that has traded between $82.75 and $89, it may not make a lot of sense to pay the premium to get the downside flexibility if prices fall unexpectedly. Other things equal, you would expect the market to go up, not down, given where it is trading in the established trading range. It might mean going directly into the futures (unless margins would for some reason be a problem), buying the futures, and pegging that specific price of $83 as compared to imposing a cost ceiling that is somewhat above that $83. You don't expect prices to go down, so why pay for the downside flexibility?

FLEXIBILITY AND PRICE DIRECTION

On the other side of the coin, the producer needs to anticipate where prices might reasonably go in deciding whether to sell the futures or buy a put option. If he is talking about forward pricing slaughter hogs and the underlying futures is trading at $59.50, the contract high is at $60.10, and the contract has traded as low as $49.80, it probably doesn't make any sense to pay the premium for a put option and buy upside price flexibility. Only two or three times in the history of the industry have cash hog prices gone above $60 per hundredweight and stayed there more than just a few weeks. Chances are, if prices move in any direction from a level that is between $59 and $60, they are likely to be down. *Sell the futures and establish the specific high price then instead of buying a put option and establishing a price floor some $1.50 to $2.00 per hundredweight typically below the specific price that you could establish.* The odds are small that the flexibility you are paying for will be needed or useful.

Don't pay option premiums if prices are not likely to move in your favor

We will come back to this consideration because many potential users of these markets will struggle trying to make the decision as to whether or not they use cash contracts, go directly into the futures, or use the options. The point that is being stressed here is that one component of the option is that you are paying a premium to buy flexibility. Don't pay the premium and buy the flexibility if the chances of that favorable price move are slim to none.

EXAMPLES CASE STUDIES: LIVESTOCK

As was the case with crops, case studies are used to clarify price risk management in the livestock commodities. Pricing both the output and inputs is demmonstrated.

CASE 1: A PRICE FLOOR FOR SLAUGHTER CATTLE

Table 8 is based on developments in the first half of 1994. Fig. 1 shows the June live cattle futures from late 1993 through June 23, the last trading day for the contract. The vertical bar shows the range of this contract price daily, and the horizontal "tic" shows the closing or settlement price for that day. This case study deals with the very difficult decision that often faces an agricultural producer who must sell as a price taker. It also demonstrates the financially devastating possibilities that can come out of the micro-macro trap discussed at length in earlier chapters.

A recent example

In late 1993 and early 1994, monthly cattle on feed reports were showing numbers on feed continuing to trend higher. Prices in late 1993 in the cash market were in the mid-to-high $70s. As Graph 3 shows, the June live cattle futures traded in a $73 to $75 range for a number of months. By late February and early March of 1994, there was increasing concern about the number of cattle that were starting to accumulate in the feedlots and the possibility of a significant supply increase in the late spring and early summer months.

Date	In Cash	In Futures
March 1	Cattle on feed are projected to reach slaughter weights by May 25. Given the cost of the 700 steers into the lot in late 1993 and estimating feed costs, the break-even price on the cattle is $78.50 per cwt.	June live cattle futures have been trading between $74-75, and closed at $74.60 on March 1. A $74 put is trading at $.90 per cwt.
March 2	Fl P = $74 - $1.00 - $1.00 = $72	Buys $74 put option on June futures at $1.00.
May 25	Sells cash cattle at $63 Net: -$15.50 per cwt. Overall Net: -$6.15 Realized Price: $72.35	June futures at $63.50, sell the $74 put for $10.35. Net: $9.35

Table 8 - Using call options to impose a cost ceiling on corn costs

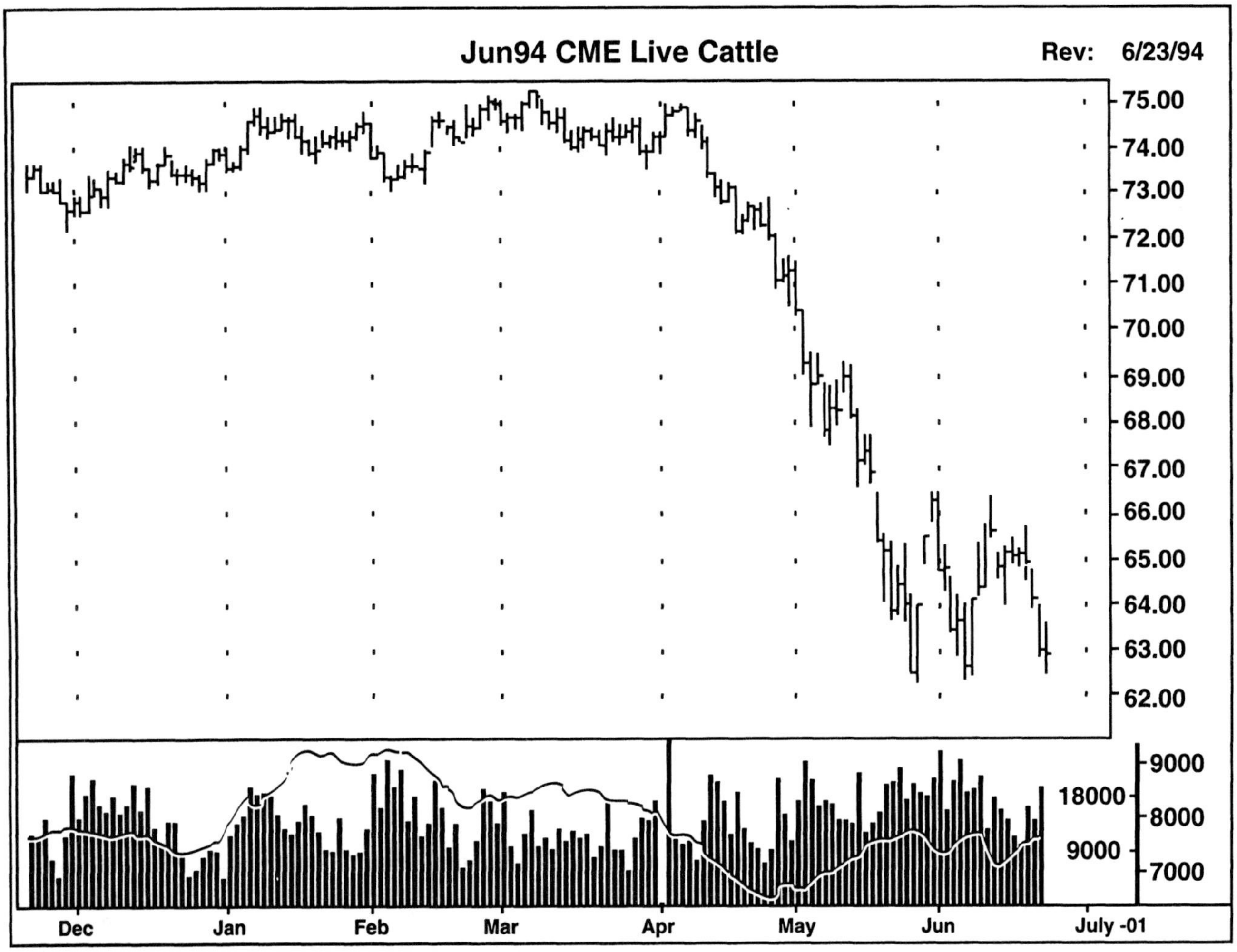

Graph 3 - Bar chart of prices for the June '94 live cattle futures

Many of the cattle that had been placed on feed and were expected to finish in the May to June period carried break-even price needs in the $78 to $80 per hundredweight range. Cattle feeders in late 1993 were optimistic that relatively favorable mid-$70s prices would get even better, and they paid higher prices for feeder cattle than, looking back, were justified. Note that the June futures was anticipating the possibility of a substantial supply of cattle. In spite of optimistic expectations, it did not trade into the high $70s, certainly not up toward $80, as many hoped it would. *It gave the typical cattle feeder no opportunity to forward price cattle placed in the last quarter of 1993 at a significant profit.*

The micro-macro trap

In this scenario, the astute producer should definitely start sniffing the air for the possibility of the micro-macro trap. In 1985 and again in 1991, the cattle feeding complex found itself in a similar situation. The futures market, anticipating supply increases, did not discover prices sufficiently high for late-spring and early-summer cattle to be forward priced at a profit. When the supply increases did start to come to market and cash prices started to weaken, there was a tendency for the individual cattle feeder to hold the cattle for another week or two waiting for prices to improve. *At this point the micro-macro trap has been sprung.* Holding cattle means they gain weight, and increased weights means less desirable cattle in terms of cutability and yield grade designations and more beef tonnage. The buying packers understand the situation, know that the cattle are backing up in the feedlots, and they start to lower their price bids. The potential for a devastating price decline is present.

HOLDING CATTLE TO EXCESSIVE WEIGHTS = **MICRO-MACRO TRAP IN OPERATION**

Watch out for the micro-macro trap!

In 1994, the placements of cattle on feed continued to increase relative to year-earlier levels through April. As the pace at which sales were being conducted declined, cattle weights started to climb. Carcass weights reached record-high levels, and beef production from only a slightly higher inventory of cattle in the feedlots climbed 7 to 8 percent compared to 1993 levels. This came in the presence of pork production that was running about even with 1993 and a 5 to 7 percent increase in poultry production. The consumers were being provided with abundant supplies of meat, and these are the conditions in which prices throughout the system usually start to soften.

The feeder's concerns

If we pick up our cattle feeder, for whom this case study is designed, on March 1, there are lots of reasons to be concerned. Waiting for the June live cattle futures to trade up toward the $80 level, so the cattle could be hedged at least at break-even, has not worked. Break-even on the cattle is $78.50 per hundredweight. The electronic market news service coming into the office at the feedlot is documenting the increase in beef production, showing the weights climbing. It also shows cutout values that are estimated daily by the USDA for boxed beef to be down in the $105 area in sharp contrast with the $125 to $130 per hundredweight levels for early 1993. The cattle feeder starts to get worried about a repeat of 1985 and 1991 and starts to think about the possibility of finding some disaster price insurance.

BUYING PUT OPTIONS BELOW BREAKEVEN SELLING PRICES = DISASTER PRICE INSURANCE

Put options can protect against disasters

On March 1, he checks his electronic system and finds that a $74 put is trading at $.90 per hundredweight. After thinking about that, he decides to go ahead and buy a $74 put option on the June futures on March 2, which ends up costing him $1.00 per hundredweight. At this point, given the -$1.00 cash-futures basis expectation, the cattle feeder has placed a price floor under his cattle at $72. That $72 is calculated by subtracting from the $74 strike price the negative $1.00 basis and the $1.00 premium cost of the $74 put. The producer has established a $72 price floor and, of course, hopes that this is one time in which price insurance will not be needed. He hopes that the market will actually trade higher into the high $70s or possibly toward the $80 level.

Graph 3 also documents what actually happened. The June live cattle futures, as cash prices started to plummet, dropped from the $75 level in early March to $62 by late May. In one 10-day period during May, the cash price dropped $7.50 per hundredweight, the largest 10-day break in the cash market in the history of the industry. The micro-macro trap obviously caught many unprotected cattle feeders. Word that of businesses who had been in cattle feeding for decades were being driven out started to circulate through the industry. Our cattle feeder finds that his cattle are ready to sell in late May, and on May 25 he negotiates a $63 selling price. Note that if no protection had been established, and this is the situation that would face the cattle feeder who is just a cash market speculator, losses would have been $15.50 per hundredweight. On a per-head basis, that means $150 to $200, and when thousands of cattle are on feed, obviously this type of price break is devastating to the cattle feeding complex.

On May 25, the June live cattle futures are trading around $63.50, which means basis on that particular day was only -$.50. The $74 put, which is within two to three days of maturity, ought to be worth essentially $74 minus the $63.50, or $10.50 per hundredweight. On calling his broker, our cattle feeder finds that the option value is not quite up to $10.50 but is very close at $10.35. He places an order to sell his $64 put on the June live cattle futures at $10.35 or better. An hour later the broker calls and indicates that they have been able to get the sale completed, and the cattle feeder has netted $9.35 from his option trade. That $9.35, of course, comes from the $10.35 value of the option less the $1.00 initial premium cost.

REALIZED PRICE OF $72.35 > CASH SPECULATOR PRICE OF $63.00

The put options would save the business from ruin

The overall net for this cattle feeder is still -$6.15 per hundredweight, which means losses as high as $75 per head. The realized price is $72.35, slightly above the floor price. This occurs because the basis was $.50 better than had been expected, but the cattle feeder had to give up $.15 of the possible maximum value on his option because there were still two or three trading days left before maturity of the option. The realized price, then, is $72.35 as compared to the floor price that would have been formerly expected of $72.

A tough decision

This poses a difficult decision for the individual cattle feeder. It is a very difficult situation for those in extension at the land-grant universities and those in private business who would presume to advise cattle feeders as to what they should do when these circumstances start to emerge. The key is to step back from the day-to-day happenings and think about what might be going on in a bigger, broader, and more aggregate context. This is, of course, the macro side of the issue. *Whether the one cattle feeder holds his cattle another week or 10 days (or even two weeks) and sees them gain additional weight doesn't make any difference.* When hundreds or even thousands of cattle feeders take the same action, it makes a lot of difference. The short-term weekly supply function of cattle that are ready to go to slaughter shifts abruptly to the right. This occurs not only because cattle movement into slaughter has not kept the needed pace but also because each animal that goes to slaughter is carrying more weight than normal. That shift in the supply function is only part of the explanation for the dramatic price decline, however.

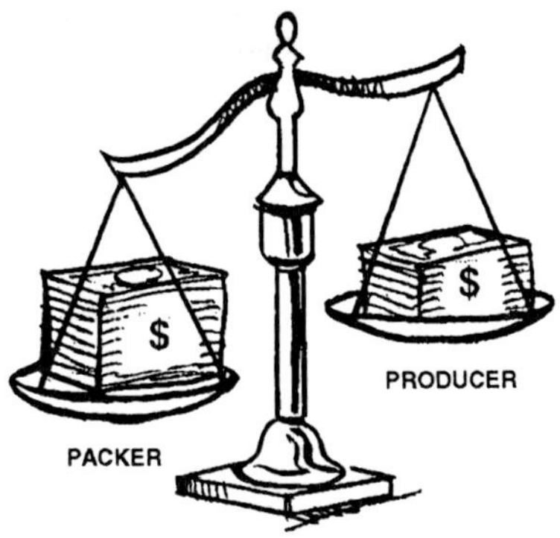

Holding cattle shifts the balance of bargaining power to the packers

Keep in mind that when the cattle feeding complex allows itself to get caught holding cattle and allows itself to move into a "must-sell" scenario, all of the advantage in the bargaining process shifts to the buyer. During the April-May-June period of 1994 when cattle prices dropped so abruptly, the margins that packers were earning were quite attractive. Retail prices, and therefore value of the product at retail, tend to come down slowly, certainly not as quickly as implied in Fig. 1 for live cattle. Thus, the value of the boxes of beef that the typical packer is moving into the service institutions or into the retail food chains stays up relative to the price he has to pay for slaughter cattle. Margins may look quite attractive during this time period, and within broad limits, the packers understandably tend to take advantage of that. After all, they may have gone through a calendar quarter in the previous year in which their margins might have been negative for several weeks or even for a few months. It is, then, a give and take situation. In this particular instance, it is the buying packer that tends to take advantage, and it is obviously the cattle feeder that "gives."

Pulling it together

There will be a "right time" to use options

This case study, as was suggested earlier in the chapter, deals with options. There are a few basic pieces of information about the option that we need to keep in mind, and they are worth collecting and summarizing here. Important dimensions to using the option would include:

- **Premiums on options tend to be highest when the markets are quite volatile.**
- **Option premiums are not only a function of volatility but also a function of time. The longer the time before expiration of the option, the higher the option premium will tend to be for given levels of market volatility.**
- **An option that is bought in early March, less than three months prior to formal maturity of the option, will tend to have a relatively low option premium.**
- **For any agricultural commodity in which physical delivery is allowed under futures contracts, as compared to cash settlement, the option matures before the last day in which the underlying futures contract trades.**
- **If there is any time left before formal maturity of the option, the option will not yet have acquired the value that it could acquire. In the case study here, a $74 put option was bought. When the futures market went down to $63.50, this technically opened up the possibility of a $10.50 value for the option. But if there are just a few days left for trade, there is still some time and some risk exposure.**

CASE 2: MANAGING COST OF CORN AS AN INPUT

This particular case is also based on a realistic scenario that prevailed in mid-June of 1994. During 1993, floods in the Midwest had pushed corn prices higher, and they remained at fairly high level into early 1994. Graph 4, however, shows that the December 1994 corn futures, the contract that would be used to forward price the 1994 crop, dropped to the $2.40's in May. Then, as planting proceeded in earnest, weather and weather forecasts turned the market very volatile. In mid-June, the December futures were around the $2.65 level, giving every impression that they could move significantly higher. A cattle feeder, hog producer, dairyman, or poultry firm that would need to buy corn all through 1994 is starting to get concerned. Recognizing that any long hedge activity or any use of call options would also probably be applied to the July, September, and December contracts, the situation presented in Table 9 deals with corn needs in the October and November period. While the felt need for protection might be mounting, the immediate question becomes one of what form should the protection take?

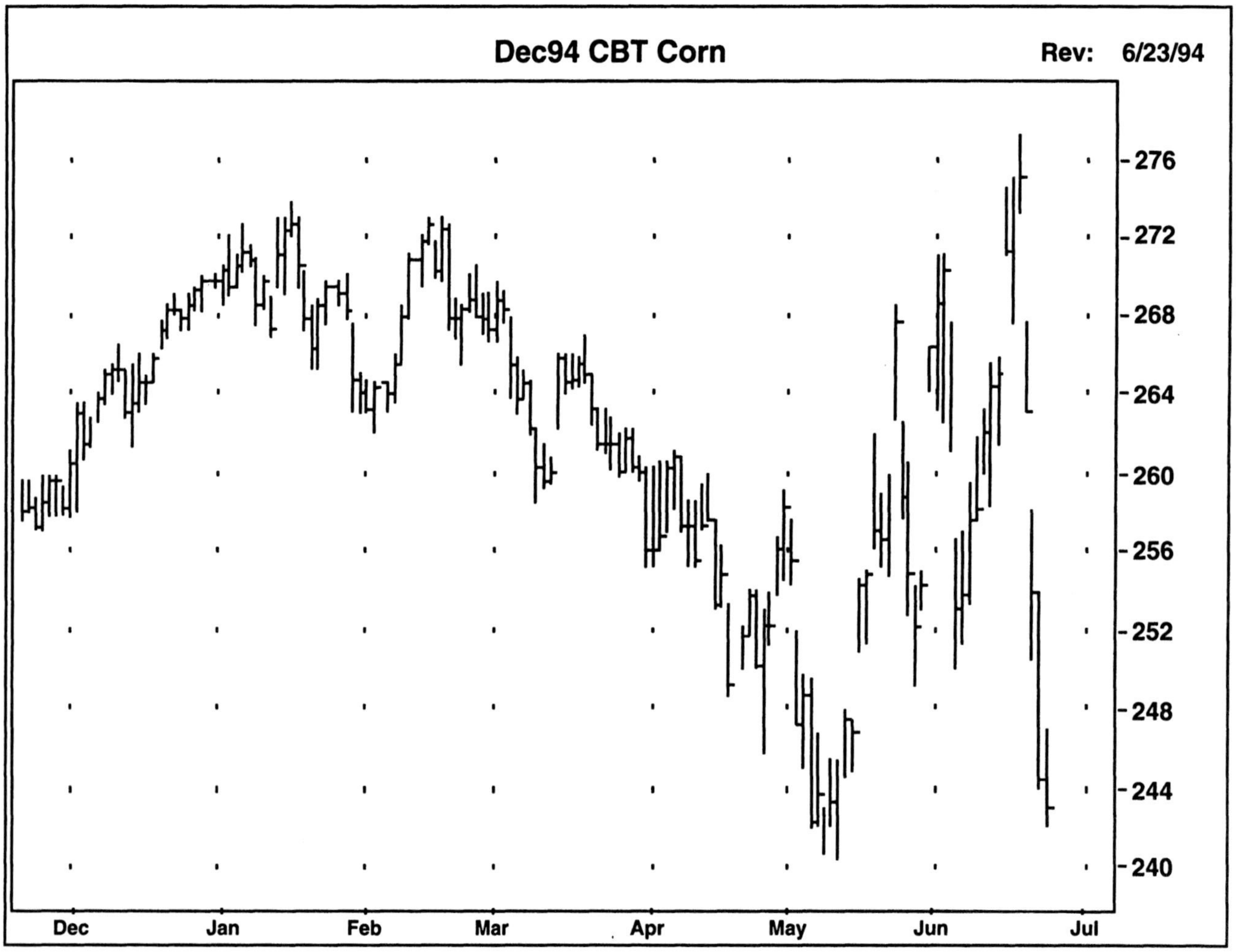

Graph 4 - Bar chart for December 1994 corn futures

Date	In Cash	In Futures
June 15	December corn futures have rallied from $2.40-2.50 in May to the $2.65 level. Weather forecasts are calling for hot and dry weather in the Midwest. You will need to buy corn in October-November.	A $2.60 call option on December futures is trading at $.21 per bushel.
June 15		Buy $2.60 calls at $.21
October 25	Cash corn prices are at $2.25 as harvest of an unexpectedly large crop is completed.	With December corn futures at $2.45, the $2.60 call has a value of only $.03.
	Net = Buy at $2.25 Net Cost = $2.25 + .18 = $2.43 per bushel	Net: -$.18

Table 9 - Using call options to impose a cost ceiling on corn costs

What form of protection?

Graph 5 helps establish some perspective. This is the long-term continuation chart for corn futures. It shows the range of the nearby futures contracts, for each month, with the horizontal "tic" the closing price on the last trading day of the month.

Note that only briefly in early 1993 have prices stayed for long below about $2.20. This suggests that if the market were at $2.40 in May, the possibility of still cheaper corn is very real but is quite small. But when the market is at $2.65, the possibility of cheaper corn, if the weather starts to improve, and the crop prospects improve is substantial. Thus, whether an

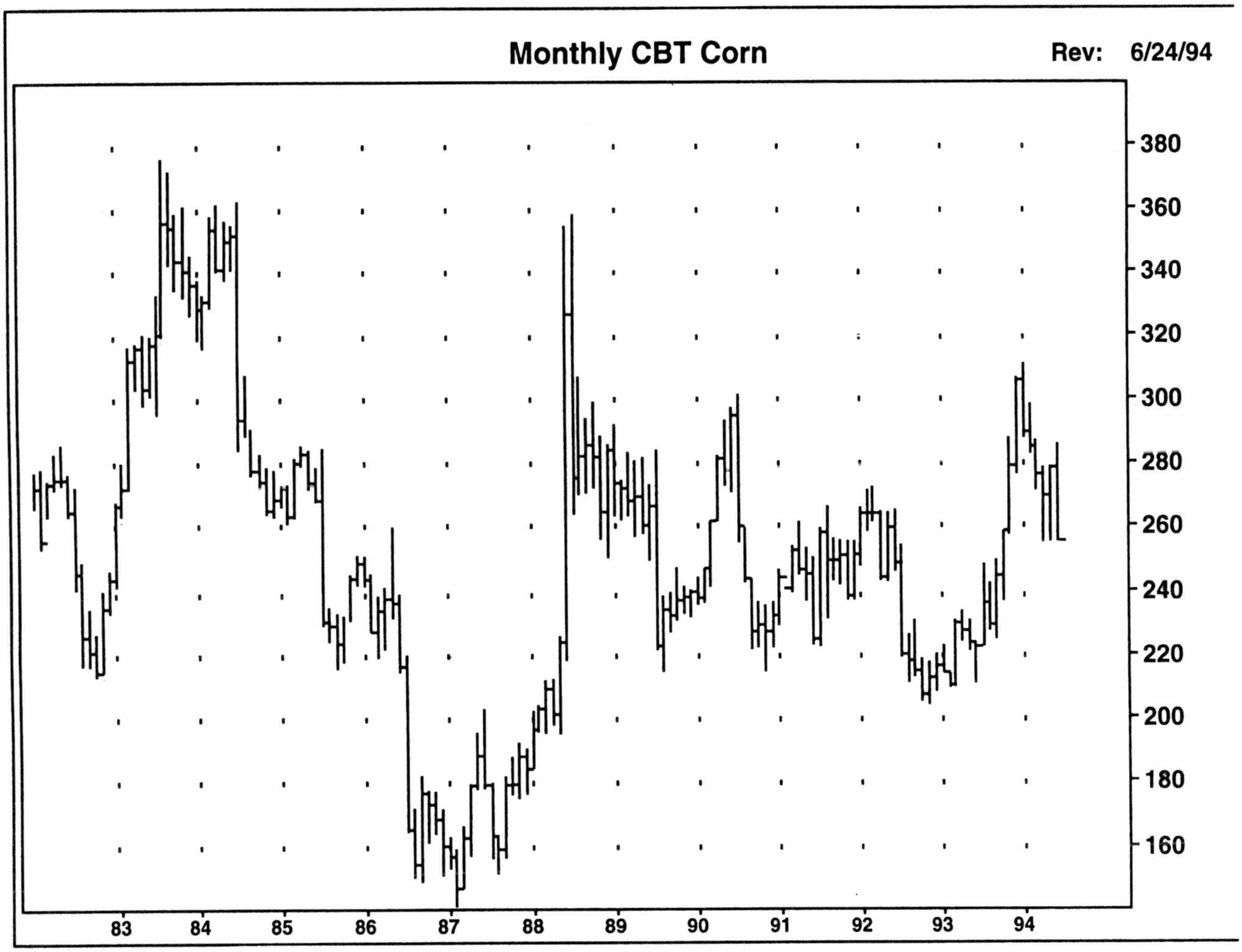

Graph 5 - Monthly continuation bar chart for corn futures

astute decision maker uses futures directly, or an option, may well depend on when this decision is being made.

If futures were bought in May when the market dipped in the $2.40's, there are reasons to argue that buying futures is the correct action. After all, paying the substantial premiums on a call option buys you the opportunity to benefit if corn costs go unexpectedly lower. When the market is already near $2.40 and history says it spends very little time below that level, the prospects for sharply cheaper corn are not very high. If anything, prices are likely to move up from that $2.40's level. If that is the case, then it is better to peg a *specific price low in the trading range for the year than it is to set a cost ceiling substantially above the specific price that is being offered.*

Table 9 presumes that the decision was not made in May and that the pressures to get some protection are mounting in mid-June. The December corn futures are up to $2.65, and a $2.60 call option costs $.21 per bushel. Given a -$.20 basis expectation in the cash market where the corn will be bought, this means the cost ceiling using a $2.60 call would be as follows:

Cost Ceiling = Futures+Basis+Option Premium
CC = SP+BASIS+ PREM
= $2.61

Applying the appropriate numbers here, we have a $2.60 strike price minus a $.20 basis plus a $.21 option premium or a cost ceiling on the corn of $2.61 per bushel. If protection is going to be established at this relatively high level in a historical context, then paying the option premiums to open up the downside potential might make economic sense.

When there is room for prices to decline, using a call option makes sense

Both Graph 4 and the bottom part of Table 11 show what can happen when the weather improves. By late June, the weather had improved and prices were again in the $2.40s. Table 9 assumes a large crop is produced. In October, cash corn prices are at $2.25, and this would mean that the December futures should be around $2.45 given the -$.20 basis expectation that had been employed earlier. With the December futures at $2.45, the right to buy those futures at $2.60, which is the $2.60 call option, is going to have very little value. It will not be zero on October 25 because this option will not expire until late November. But it will have a very low value. In this instance, we assume we have little time left, and in a $2.45 market, the option only has a value of $.03. In the cash, your net is $2.25 since you can buy cash corn at $2.25. In the options trade, your net is -$.18 per bushel. You paid $.21 per bushel for the cost protection and were able to sell that right, or that option, at only $.03 per bushel in late October; thus a net of -$.18 per bushel. The net cost of the corn, then, is $2.25 plus $.18 or $2.43 per bushel.

Net costs

Note that the cost is above the cash market by the net cost of the option. Note also that this producer would be facing somewhat higher costs going into a dairy or livestock program than would the cash market speculator in this particular year who could buy corn at $2.25. It is also interesting to compare the net cost of the corn of $2.43 under the option program with what the net would be if the futures had been bought at the $2.65 trading level on June 15. Given the $.20 basis, the specific price would have been $2.45, leaving very little difference between the net cost in that particular instance and the net cost from the option program.

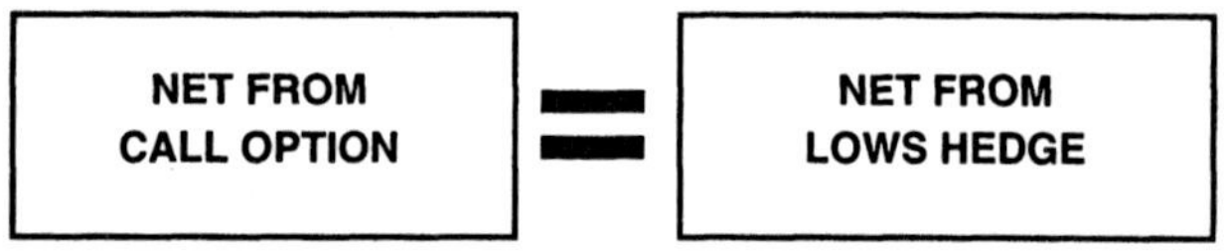

Options and long hedges net about the same cost, but options offer added advantages

But that net cost is not necessarily the appropriate thing to use in comparing the two approaches in mid-June. Given that the weather patterns and the forecasts were, in May, starting to project sharp reductions in yields and significantly higher corn costs, buying the option might make more sense because the decision is being made when the futures are up to $2.65. From a $2.65 price plane there is substantial room for costs to get cheaper in the event the supplies are adequate. Thus, it might make more sense to go ahead and use the call option and eliminate any margin considerations in futures, etc. In fact, that is the decision that most astute decision makers would make in this particular instance. The only time that one might argue that the futures should be bought directly is if this decision had been made more appropriately in hindsight. But hindsight is always 20:20, is it not? This decision maker had to act before all the weather variables had worked themselves out.

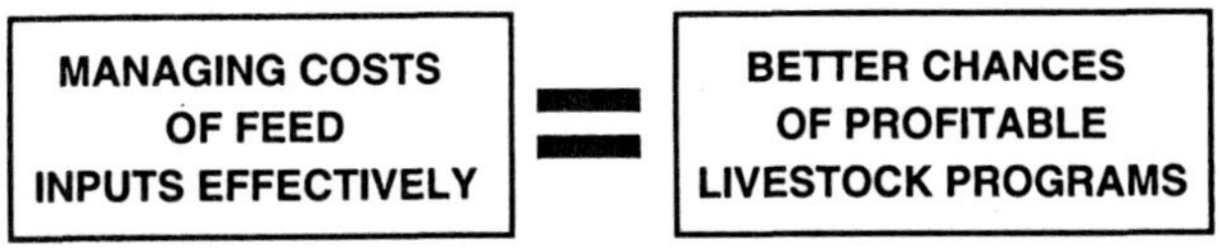

Managing feed input costs is critical to profitability

SUMMARY

Procedures for hedging livestock commodities are very similar to those demonstrated for grains in Chapter 6. Basis is still involved, and margins and margin calls are still involved.

The livestock commodities can sometimes show a variable basis, especially in feeder cattle. The lack of stability or predictability of the closing basis means the basis risk can be significant. Quality and weights can vary in the cash market, but the futures contracts are always written for a specific quality, sex of cattle, and weight range.

The case studies demonstrated the vagaries of weather and the *micro-macro trap* which had been discussed in detail in earlier chapters, especially Chapter 2. For livestock finishing programs, the costs of feed inputs are very important. Using call options to place a ceiling on corn costs, the discussion was extended to when to use call options versus long positions in futures:

- **When there is reason to expect corn costs might decrease.**
- **When futures prices are well above the bottom end of the trading range for the year.**

Using puts or calls in feeder cattle or any livestock program will often make sense. As suggested, the uncertainty from the corn and feedgrain markets filters into and affects prices, making livestock prices doubly volatile. The micro-macro trap in cattle feeding was demonstrated to be very dangerous when cattle are held off the market hoping for price improvement. Excess weights can result, leading to disastrous price declines that can be hard to manage.

8

Analyzing the Markets

INTRODUCTION

In earlier chapters, there has been considerable discussion about price movement over time and the risk and uncertainty associated with those moves. In grains, weather can be a major factor. In the livestock complex, seasonal and cyclical surges in supply can create price problems for the producer. When we step back and remember that the producer is a price taker, it becomes clear that there is a major need to protect against disastrous prices. Getting that protection requires forward pricing. Forward pricing techniques have been introduced. We can use cash contracts, go directly into the futures, or use options on futures. But the lingering questions are, "*Which way are the markets going? Is now the right time to forward price? Should I use futures, options, or cash contracts?*"

In this chapter, we look at some very basic analytical tools to help the decision maker decipher what is going on in the marketplace and answer those questions. A quick look at the literature in this area will indicate there are two broad approaches to analysis of the markets. The first is called a **fundamental approach**, and this approach deals with the basic economic forces of supply and demand that have already been introduced and discussed. The second is a **technical approach**. This type of analysis relies on the past pattern in prices for a particular commodity to tell something about where future prices will be.

TECHNICAL AND FUNDAMENTAL ANALYSIS CAN BE COMPLEMENTARY

We can work together

Sometimes the two approaches are treated as if they are competitive. They need not be. You can use both approaches to analyzing the markets. In particular, it makes sense to apply a fundamental approach to the market to determine the likely *direction* of prices from any one point in time. Then, the technical dimensions of the market can be used to help reinforce that earlier impression on direction. The technical approaches help mostly along another dimension, however. They can definitely be used in helping decide the *timing* of entry into the markets. There is no reason why both approaches cannot be used on the same problem. Both will, in fact, be discussed in this chapter.

APPROACHES TO MARKET ANALYSIS

Here we will deal first with the fundamental approach for the grain and livestock commodities. Emphasis will be on very basic measures and on the type of information that decision makers need to have at their disposal. The important consideration is always one of trying to determine which way a price will move. The fundamental tools will be presented with that thought in mind, and some care will be taken in showing how each particular approach or each particular piece of the analysis can be used to determine which way prices are likely to move. After that is completed, we will introduce price charts. There are a few reliable, very simple technical indicators of market performance, and they will be highlighted as tools for the decision maker.

These supply-demand reports are very helpful in keeping up with the markets

FUNDAMENTAL ANALYSIS

In the grains, the key base of information comes from the USDA supply and demand reports. Since the early 1970's, the USDA has been required to issue several times a year their estimates of what the supply of grains, oilseeds, cotton, etc., are likely to be and their estimates of the demand for those commodities. The demand component includes both domestic and export. In the past two decades, the export demand has been very important. In wheat, for example, roughly 50 percent of the crop is normally exported. In some years that level has reached 60 to 70 percent. Obviously, then, what is going on in the world market as it influences export demand for our commodities will be an important price determinant.

Item	1992/93	1993/94	1994/95 Projections	
			May Est.	June
		Million acres		
Planted	79.3	73.3	78.6	78.6
Harvested	72.2	63.0	71.5	71.5
		Bushels		
Yield per harvested acre	131.4	100.7	122.1	122.1
		Million bushels		
Beginning stocks	1,100	2,113	827	832
Production	9,482	6,344	8,725	8,725
Imports	7	25	5	5
Supply, total	10,589	8,482	9,557	9,562
Feed and residual	5,301	4,825	5,200	5,200
Food, seed & industrial	1,511	1,600	1,750	1,750
Domestic, total	6,813	6,425	6,950	6,950
Exports	1,663	1,225	1,350	1,350
Use, total	8,476	7,650	8,300	8,300
Ending stocks, total	2,113	832	1,257	1,262
Farmer-owned reserve	13	100		
CCC inventory	56	45		
Free stocks	2,044	687		
Outstanding loans	693	125		
Avg. Farm price ($/bu)	2.07	2.50-2.60	2.10-2.50	2.10-2.50

Table 1-U.S. corn supply and use

Table 1 shows a sample supply-demand report for corn. The basic format is the same for corn, wheat, cotton, soybeans, etc., and will not be repeated. These reports are available from the USDA at a nominal cost, and directions on how they can be ordered are included in the appendix .

Note the format in the table is something of an accounting format. There is always a beginning stocks level to which the production in any particular crop year is added to give a total supply. Then various components of demand are identified. In the case of corn, the big component of demand is domestic feed use. (It is shown as "feed and residual," but the residual is an accounting need and is of minor importance.) There are also some domestic uses for non-feed purposes. Export demand is also important. By adding up the components of demand and subtracting those from the total supply, we get an estimate of ending stocks for the particular crop year.

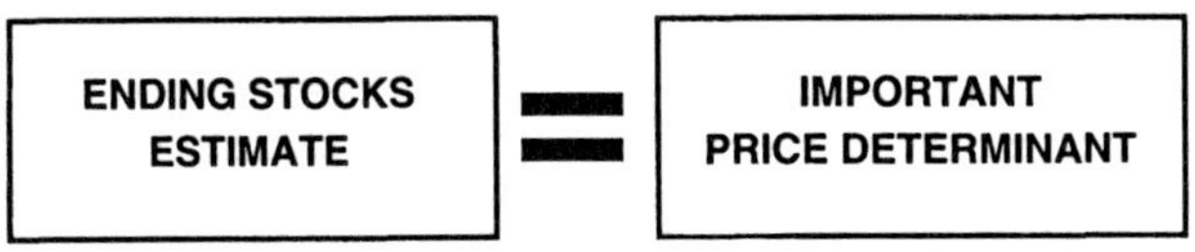

Traders watch ending stocks carefully

Ending stock

Perhaps the most important number in the table is the ending stocks number. There are two components of ending stocks that market traders watch carefully. The first and perhaps most important is the level of stocks. *How large the ending stocks are is a direct measure of the "buffer" or protection that the markets have against weather or other problems that would reduce supplies in coming growing seasons.*

Ending stocks are a wall of protection against weather problems

When this buffer is very large, prices not only tend to be low, they also tend to be relatively stable. What the marketplace is telling us is that when we have two or three billion bushels of corn left over, for example, (or even more in some years) then there is no reason to get excited about short-term weather problems that might influence yields and the crop size in any particular year.

Marketing Year (beginning Sept. 1)	Ending Stocks (mil. bu.)	Average Farm Price ($/bu.)
1975/76	633	2.54
1976/77	1,135	2.15
1977/78	1,436	2.02
1978/79	1,710	2.25
1979/80	2,034	2.48
1980/81	1,392	3.12
1981/82	2,537	2.47
1982/83	3,523	2.55
1983/84	1,006	3.21
1984/85	1,648	2.63
1985/86	4,040	2.23
1986/87	4,882	1.50
1987/88	4,259	1.94
1988/89	1,930	2.54
1989/90	1,344	2.36
1990/91	1,521	2.28
1991/92	1,100	2.37
1992/93	2,113	2.07
1993/94	850	2.50
1994/95	1,930*	2.15*
*Estimated		

Table 2-Ending stocks for corn and average farm price, 1975-1994

This is all quite logical and makes sense. Table 2 records the ending stocks by crop years for corn and average prices for those particular years. It is not at all difficult to examine these numbers and see that when the stocks are large, price tends to be low. When stocks are relatively small, price tends to be high.

The other dimension of ending stocks that is important is direction of movement. Obviously, when the buffer stocks are going down from one year to the next, there is an increased possibility of supply-side problems and shortages. Thus, when the direction of movement in stocks is down, other things equal, the market tends to discover higher prices. This will be especially true when the direction of stocks is down and the level of stocks is also relatively small.

Price and ending stocks

Graph 1 is perhaps the best way to demonstrate what is happening. It shows average prices of corn received by farmers across a number of recent crop years on the vertical axis and, on the horizontal axis, ending stocks as a percent of use for that year. This horizontal measure is a representation of how large the stocks are relative to usage. As you move to the left and toward the origin down to, say, 15 percent, that means the stocks are getting relatively "tight." The buffer against weather problems is small. If you move in the other direction toward 30 or 40 percent, that means the stocks are very abundant compared to a normal year's usage. What you see on the graph is that when the stocks relative to usage are moving to the left, or getting smaller, prices tend to be higher and vice versa.

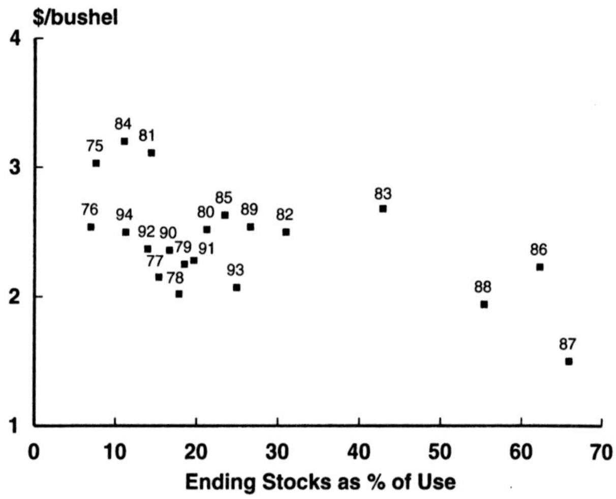

Graph 1 - Price vs. ending stocks as a percent of use: corn

This type of simple plot is a very useful indicator of the relationship between supply, as measured by the available carryover stocks, usage and price. It will give any observer some perspective of the markets and help them to understand why prices are volatile when stocks are relatively low and why they are stable when stocks are quite large. The years shown in the chart are the last years in a crop year. For example, 94 refers to the price/ending stock coordinates for the 1993/94 crop year which started September 1, 1993, and ends August 31, 1994. The ending stocks figures in the supply-demand table for 1993/94 are the stocks levels on August 31, 1994.

Graphs 2 and 3 show a similar plot for soybeans and wheat respectively. (The crop year for soybeans is September 1 to August 31; for wheat it is June 1 to May 31.) The general relationship is there in both commodities. As stocks relative to normal usage get small, price tends to be high. As the stocks relative to normal usage get larger, prices tend to be lower. Note that the relationship between price and stocks on all three figures does not necessarily look like it would be a straight line. When stocks get very small in any of the commodities, price bursts to the upside. What might best fit, if you drew a line through these points to represent them, would be a function that has some curve to it and starts to flatten out as the stocks get larger.

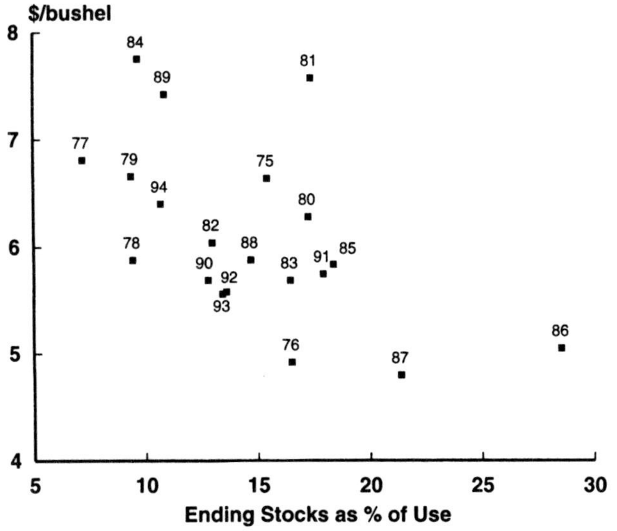

Graph 2 - Price vs. ending stocks as a percent of use: soybeans

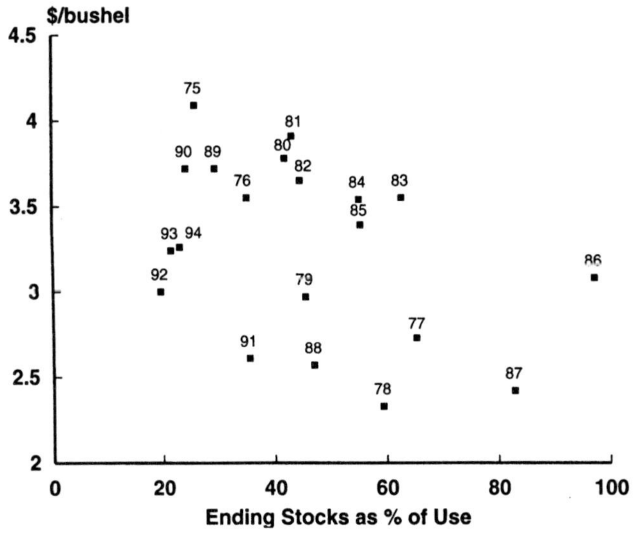

Graph 3 - Price vs. ending stocks as a percent of use: wheat

Prospective plantings

A producer who monitors the supply-demand reports that are released monthly during the spring and summer months, and perhaps somewhat less frequently in the winter months, has a base for understanding something about what is going on in the markets. Early in the year, there are other USDA reports that will start setting the tone for the year. For example, there is the very important *Prospective Plantings* report that is released before the corn is planted. It was last released on March 31, 1994, for example. This shows how many acres producers say, in response to a survey, they are going to plant to corn, soybeans, wheat, cotton, peanuts, etc. *This is very important information. The USDA can then take that estimate of acreage, apply a projected yield to it that assumes normal weather, and get an early and initial estimate of production for the year.* By adding that to the carry-in or beginning stocks, we start to get some notion as to what the total supply is going to be for the year. That estimate of total supply obviously starts to establish one very important side of the price equation. The things that are still unknown, of course, is what weather will do in other producing countries, how aggressively importing countries will buy, and therefore what the level of export demand will be. That is what makes these markets interesting and challenging!

The prospective plantings report is the first important market report of the year

By starting early in the year, then, with the *Prospective Plantings* report and monitoring supply-demand reports during the year, the producer should be in a position to do some basic fundamental analysis and understand the direction in which prices might move. There are a number of market advisory services that offer, for a nominal annual fee, their interpretation of all these USDA data and their recommendations as to what producers should be doing in terms of forward pricing. Producers, then, basically have a choice: either be reasonably well prepared in terms of analyzing market behavior on their own or buy that advice from an advisory service. It is not a bad idea to be fairly well informed on your own and to buy the services of a good market advisory service so that you can more fully and completely understand what is being recommended and why. Widely published magazines that offer access to various advisory services are listed in the appendix at the end of this chapter. Your extension marketing specialist at the land grant university will also know about some of these advisory services.

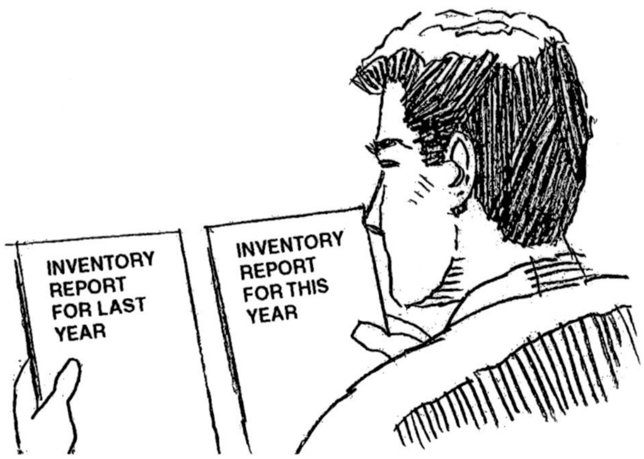

The January inventory reports are early indicators of livestock supply for the year

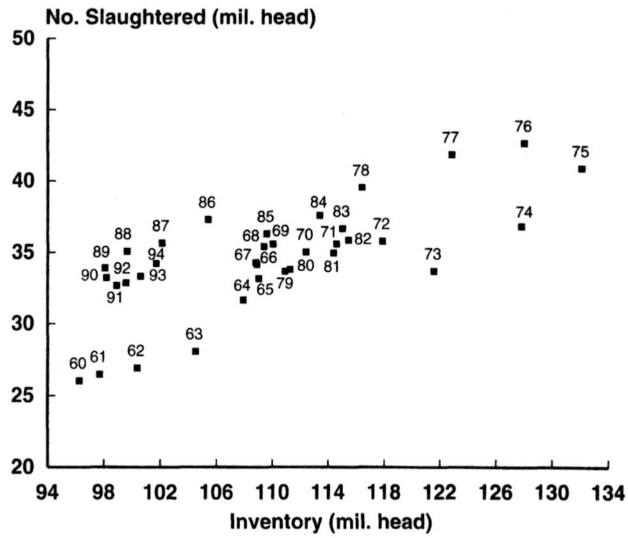

Graph 4 - Commercial slaughter of cattle and calves and January 1 inventory, 1960-94

SUPPLY IN LIVESTOCK

In the livestock commodities, the key pieces of information come from yearly inventory reports and then, periodically within the year, indications of supply. If we look at cattle first, the most important initial piece of information is the January 1 inventory on cattle numbers, typically released in late January. This sets the base from which slaughter, and therefore supply, within the year must be pulled. There will be a strong relationship between inventory numbers and slaughter except in those years in which cyclical liquidation or expansion is so strong that they distort the numbers. Graph 4, for example, plots slaughter on the vertical axis, against January 1 inventory numbers at the beginning of the year, on the horizontal axis. The relationship is not precise, of course, but there is clearly a tendency for slaughter to be higher during those years in which the inventory is high and vice versa.

Cattle on feed

During the year, the important reports are the *Cattle on Feed* reports. States included have varied somewhat over time, but currently there is a quarterly report from 13 major feeding states and a monthly report from 7 of the largest feeding states. Table 3 shows a format for one of the monthly reports. Information provided includes placements into the feedlots during the prior month, marketings from the feedlots during the prior month, and a cattle-on-feed count both at the beginning of the month and at the end. Usually, two or three years of data are shown in the reports, as is the case in this table, just to allow some simple comparisons to year-earlier and two-year earlier levels.

Item	Number			1994 as % of	
	1992	1993	1994	1992	1993
	1,000 Head			Percent	
On Feed May 1[1]	7,818	8,329	8,591	110	103
Placed on Feed During May	1,724	1,791	1,405	81	78
Fed Cattle Marketed During May	1,594	1,666	1,679	105	101
Other Disappearance During May[2]	122	131	92	75	70
On Feed June 1[1]	7,826	8,323	8,225	105	99

1 Cattle and calves on feed are animals for slaughter market being fed a ration of grain or other concentrates and are expected to produce a carcass that will grade Select or better.
2 Includes death losses, movement from feedlots to pastures and shipments to other feedlots for further feeding

Table 3 - Cattle on feed, placements, marketings, and other disappearance, 7 states, June 1, 1992-94

In interpreting this table one might recall the earlier discussion about the supply and demand reports for grains. Behind the scenes, there is some level of demand that can be represented by a downward sloping demand curve for beef. What these reports are doing is helping to formulate an impression of where the supply is and where you will get an intersection or a match between supply and demand. *This intersection generates the market-clearing price.* If, during a particular month, placements into the feedlot exceed marketings by a substantial margin, then the on-feed count will increase during the month. Some four to five months later, depending on the weight of the cattle when they go into the feedlot, these cattle will reach slaughter weights. *It is not difficult, then, to recognize that when the cattle-on-feed count is growing significantly, the supply of beef to come to market some months later is also going to increase.* If the on-feed count is surprisingly high, the price discovery process moves into action. The trader of live cattle futures at the Chicago Mercantile Exchange shifts his or her expectation of the supply function to the right compared to earlier expectations. In the process, all traders (speculators and hedgers) lower price expectations and tend to go into the trading pits sell or buy futures accordingly. This is the price discovery process at work, and these *Cattle on Feed* reports are of basic importance in determining prices for cattle.

NUMBER OF CATTLE IN THE FEEDLOTS = **SHORT RUN SUPPLY MEASURE**

The cattle-on-feed number sets the short run supply

Hog and pig reports

In hogs, reports are quarterly. The format is shown in Table 4. Basically, we get supply information that is somewhat parallel to what is offered in the cattle sector. A very useful addition here is how many gilts are moved into breeding herd. This is important to the short-term and long-term price outlook for hogs. Thus, the reports show total numbers, the numbers being held for market, and the numbers being held for breeding.

When the breeding herd number is several percentage points above year-ago levels, the astute producer has the basics that are needed to formulate price expectations, at least in terms of direction. If the breeding herd is up 5 percent, for example, and that translates into something approaching a 5 percent increase in supply some 3 calendar quarters later, it seems likely that price in that quarter will be down compared to year-ago levels. *It is this type of basic economic reasoning that will be important to the producer who is trying to decide on direction of price movement and what price level might reasonably be attained in future weeks and months.*

DEMAND FOR BEEF, PORK

The demand side may be easier to monitor in livestock and meats than it is in oilseeds. Remember that in grain and oilseeds, it is not only domestic demand but also export demand that you have to monitor. Watching how those estimates are changed by the USDA in periodic reports helps you understand what is happening to demand. Most observers just watch the supply-demand reports, the ending stocks, and use the graph of price against ending stocks as their primary frame of reference.

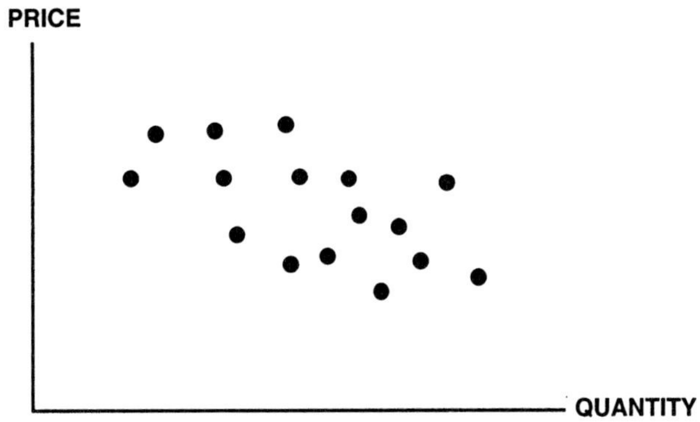

A scatter plot of price and quantity can help identify demand patterns

In livestock and meats, scatter plots for price and quantity are more widely used. Those scatter plots are presented below for beef, pork, and broilers, but it is useful to establish some sort of frame of reference before we start interpreting what the plots mean.

Item	Number 1992	1993	1994	1994 as % of 1992	1993
	1,000 Head			Percent	
March 1 Inventory					
All Hogs and Pigs	56,190	56,485	55,630	99	98
Kept for Breeding	7,155	7,050	7,025	98	100
Market	49,035	49,435	48,605	99	98

Table 4 - Hogs and pigs inventory numbers, United States, 1992-94

Graph 5 shows the possibility of a starting point at some particular level of price and quantity, labeled R for reference point. Around that reference point the diagram is divided into four quadrants labeled A, B, C, and D. This graphical setup is useful to determine what is happening to demand for a particular commodity. Remember, a demand curve and a supply curve intersected at point R. This is how the market-clearing price gets established. As we move ahead, it would be useful for the reader to visualize a negatively sloping demand curve passing through point R as it intersects with a positively sloping supply curve.

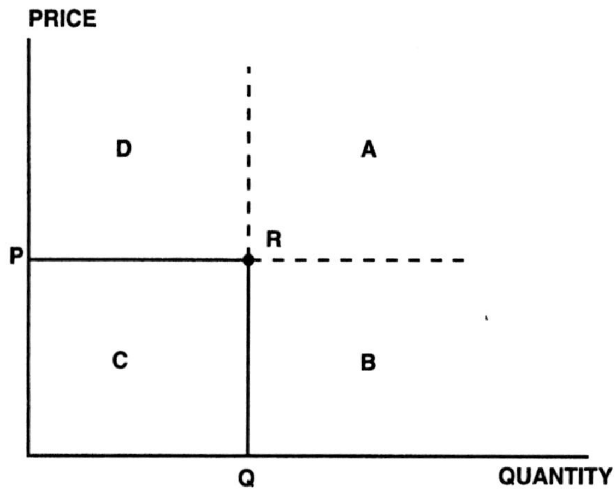

Graph 5 - A Framework For Interpreting Changes In Price-Quantity Plots

Graph 6 shows inflation-adjusted or deflated prices for Choice beef and per capita consumption as a measure of quantity or supply since 1960. The numbers in the graph indicate the years. Given this scatter plot, it is possible for the reader to determine something about what has been happening to demand for beef in recent years. Consider, for example, 1979 as the price-quantity combination comparable to our reference point R above. Note that from 1979 to 1986, the per capita supplies and therefore per capita consumption were essentially constant around 77 to 78 pounds, and price declined by over 30 percent. From 1987 through 1993, prices were more nearly maintained, in inflation-adjusted terms, at a roughly constant level but only because quantities offered continued to decline. In 1994, per capita supplies (and consumption) increased for the first time in many years and the price declined sharply. *Moving down into this quadrant from the initial reference point R is clearly a decrease in demand.* The reader is encouraged to visualize a negatively sloping demand curve passing through the coordinate at 1979 in Graph 6 and then through the coordinates labeled 1993 and/or 1994 in that same figure. Clearly, demand in 1993 was well below and to the left of demand in 1979.

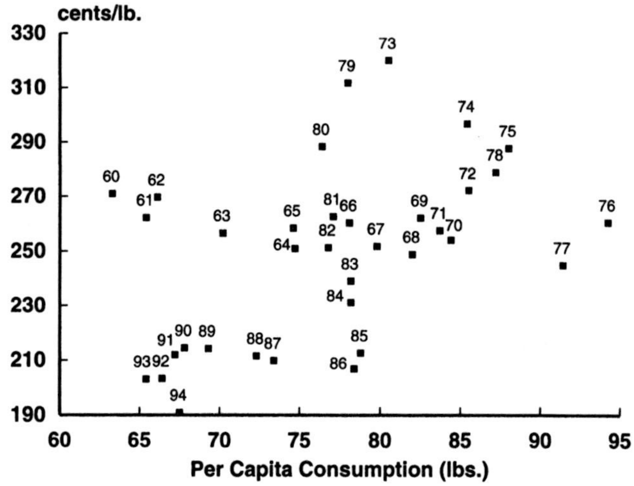

Graph 6 - Inflation-adjusted prices (CPI, 1982-84=100) and per capita consumption for beef, 1960-94

Adjusting for price inflation

Before moving on, let's briefly discuss what **"inflation-adjusted prices"** are and what this means. To get at what is happening to price, quantity, and demand (which involves price *and* quantity) over time, we need to remove the tendency for prices to rise just because of price inflation. Using the CPI to "deflate" the price data gets the influence of price inflation out and allows us to make legitimate comparisons over time.

This is very important to cattle producers. If demand for beef at the consumer level declines, consumers will take the same quantity only at a lower price. This is what we saw happen in the marketplace from 1979 through 1986. Alternatively, they will take lesser quantities if price stays essentially the same, and this is the pattern that was observed from 1987 through 1993. *If the only way you can sell even a decreased quantity of product is at a lower price (and that is generally what is happening as you move into this particular quadrant from the initial reference point), then something is wrong on the demand side.*

Something is wrong on the demand side

During the time period we have been discussing, from 1975 into the mid-1990's, decreases in demand for beef appear to have been very significant. They are related to fat content in the product, perceptions about cholesterol, and, perhaps very importantly, to the lack of convenience in preparing beef for meals in the modern household. Those declines in demand have had a serious negative impact on the industry. From 1975 into the early 1990's, the total inventory number of all cattle and calves declined from over 132 million head to 98 million-head. That same period showed the beef cow herd in the United States moving from over 47 million head down to 33 million-head, forcing the exit of roughly 300,000 average-size beef cow producers in the United States. *Obviously what is happening on the demand side is important, and the astute decision makers will need to factor this into their thinking.*

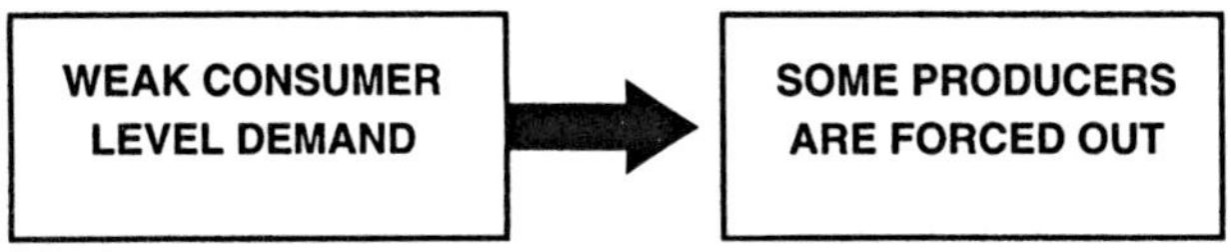

What happens on the demand side is critical to producers

There are two obvious decisions that are influenced by what is happening on the demand side. The first is whether or not it makes sense to expand cattle production or might investment funds be redirected to some other use. In the shorter run, and more nearly consistent with the orientation of this book, all this is important to marketing decisions. The case study in Chapter 7 documented the dramatic price declines that occurred in the first half of 1994, coming primarily from a surge in supply as cattle were held in the feedlots to excessive weights. But that price decline would have been less severe if there had been strong and growing demand for beef during that same six-month period. There is every indication that the pattern observed in Graph 6 is continuing into 1995. Producers did not get help from the demand side when the supplies increased (especially in 1994) and started to push prices lower. What this all means is that one should keep in mind what is happening to the demand side for beef while watching the *Cattle On Feed* and other supply-oriented reports in formulating price expectations.

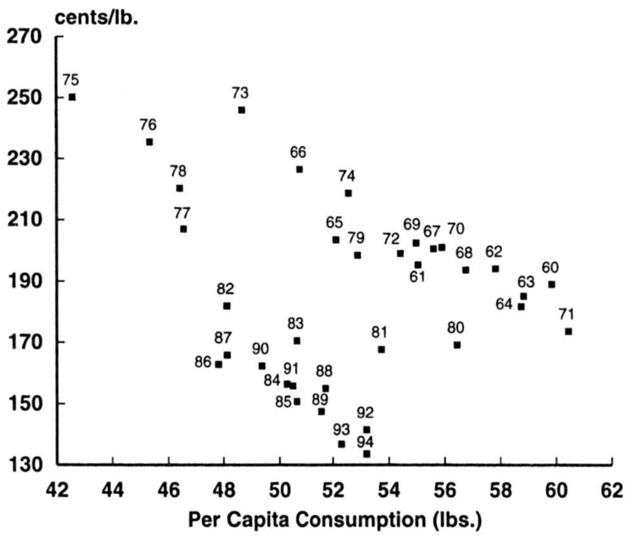

Graph 7 - Inflation-adjusted price and per capita consumption for pork, 1960-1994

Graph 7 shows a scatter plot for pork. There appear to have been demand problems in pork as well as beef, but not as severe and not as sustained. The reader might think about using the 1980 coordinate as a reference point here. Note that there is a general tendency to move to the left and down since that time period. The lowest level of demand may have been reached in 1986, and there are some year-to-year indications of improvements since that time. The 1994 observation does not look very positive, however, and it appears that in the presence of continuing weakness in beef, and increased supplies and increased competition from poultry, the demand for pork is not as strong as a hog producer might like it to be. Thus, what is happening on the demand side in pork needs to be kept in mind as the *Hogs and Pigs* reports are monitored and price expectations are pulled from those measures of supply, and potential supply.

Graph 8 and 9 show two different scatter plots for broilers. The pattern here is significantly different, and it is interesting to reflect on why.

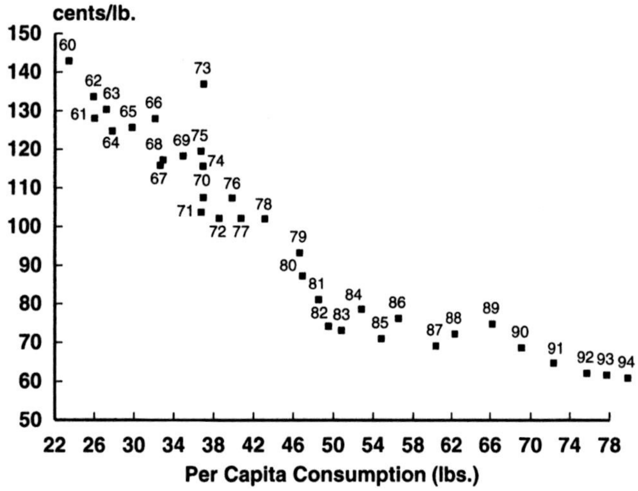

Graph 8 - Inflation-adjusted price and per capita consumption of whole broilers, 1960-1994

Graph 8 shows the deflated or inflation-adjusted prices for whole broilers plotted against per capita consumption going back to 1960. The reader is encouraged to use 1980 as a possible reference point here. Compare, for example, 1989 with 1980, and the inevitable conclusion is that a move has occurred up and into the demand increasing quadrant, quadrant A, back in Graph 5. Clearly, 1989 would have been on a significantly higher negatively sloping demand curve than would 1980. Clearly, over that time period, the poultry firms were able to sell more product at higher inflation-adjusted prices. That is a textbook indication of increases in demand. Since 1989 and through 1994, the situation stabilized somewhat. It may be that most of those years are falling near a common negatively sloping demand surface. Typically, when demand increases over time, it tends to surge for a few years, then stabilize, then surge again. It is possible that the poultry industry will see more demand increases later in the decade.

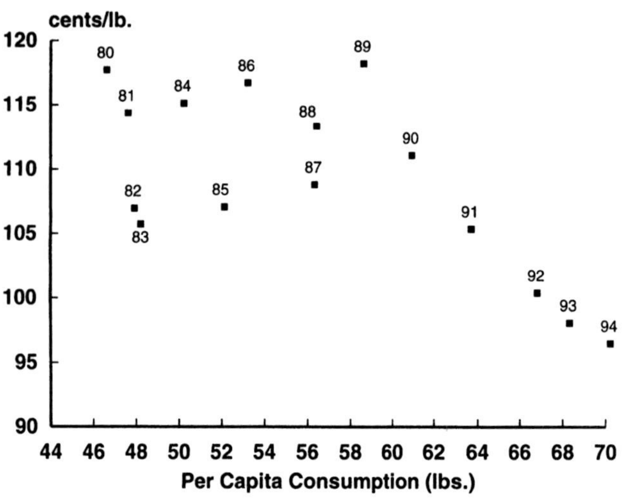

Graph 9 - Inflation-adjusted price and per capita consumption for retail broilers, 1980-1994

Graph 9 shows a plot that goes back only to 1980. This plot uses a price series that is based on quantities bought in the form of whole broilers, thighs, and breasts (where much of the consumption now resides). In other words, the price is a weighted average price using quantities of breasts, thighs, and whole birds as the "weights" on prices. This price pattern looks even more impressive from 1980 through 1989 than did the one for whole birds. The pattern from 1989 to 1994 looks similar as the industry continues to be able to offer more product at lower inflation-adjusted prices, which is an indirect indication of cost-reducing technology being applied.

Competition for the consumer's food dollar

Clearly, Graphs 8 and 9 suggest that the broiler industry is a growth industry in sharp contrast to beef. Per capita production and supply, and therefore per capita consumption, are increasing for broilers. The increase based on cost-reducing technology, but very importantly during the 1980's, also on substantial increases in demand. *This type of analysis of what is happening in poultry is important not only to the poultry firms but also to beef and pork producers as they think about increasing competition for the consumer market.*

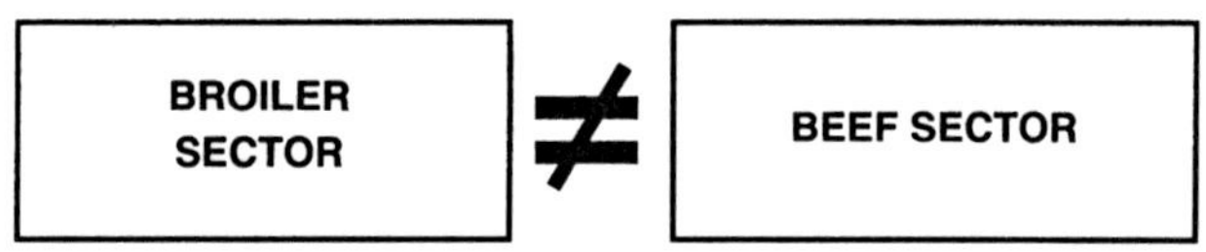

Broilers are in a growth mode; beef is not

In fundamental analysis, then, it is important to monitor developments in supply, but demand should not be overlooked. It is shifts in supply that prompt most of the within-year price moves, but demand dominates in determining price direction and price levels in the long run. If demand is weak or declining over time, the supplies *will* eventually decrease unless costs can continually be reduced to keep producers in business. This is not always possible. The beef industry provides an example of demand-prompted reductions in price that were too big to be overcome by becoming more efficient.

TECHNICAL ANALYSIS

Technical analysis is based on the idea that the past pattern of prices has some power in predicting future prices. The approach has its critics and its supporters. Some people argue that since the information flows that tend to hit the markets are largely random, then day-to-day changes in the discovered prices should be independent of each other. Others argue quite the opposite, suggesting that trading behavior in the marketplace and price expectations in the marketplace are formed not only by fundamental considerations but also by, for example, the direction of price trend shown on the chart.

Technical analysis helps with the timing of pricing decisions

A debate over who is right and wrong is not productive. Long experience by the author suggests that technical analysis is a very convenient way to monitor what is going on in the market. It helps you decipher responses to changes in fundamental information, and helps you determine when its time to "pull the trigger" and establish a forward price. This latter point, the timing of pricing actions, is probably the contribution from technical analysis that most market analysts would agree is worthwhile.

Entire books have been written on this subject. Some useful references are listed in the appendix. There will be no attempt in this book to replicate what a much bigger literature encompasses. The idea here is to focus on basic technical considerations that will help the decision maker who wants to become a better marketer develop and execute a marketing plan. Although there are various ways of charting the market and analyzing its behavior, the most widely used is the bar chart. Coverage in this section will be restricted to comments on the bar chart and how to use it.

BAR CHART OF COMMODITY PRICES = **MOST WIDELY USED PRICE CHART**

The bar chart is the preferred market analysis tool

Bar charts

Graph 10 shows a bar chart for wheat. The intent here is to help the reader understand what is involved and how the

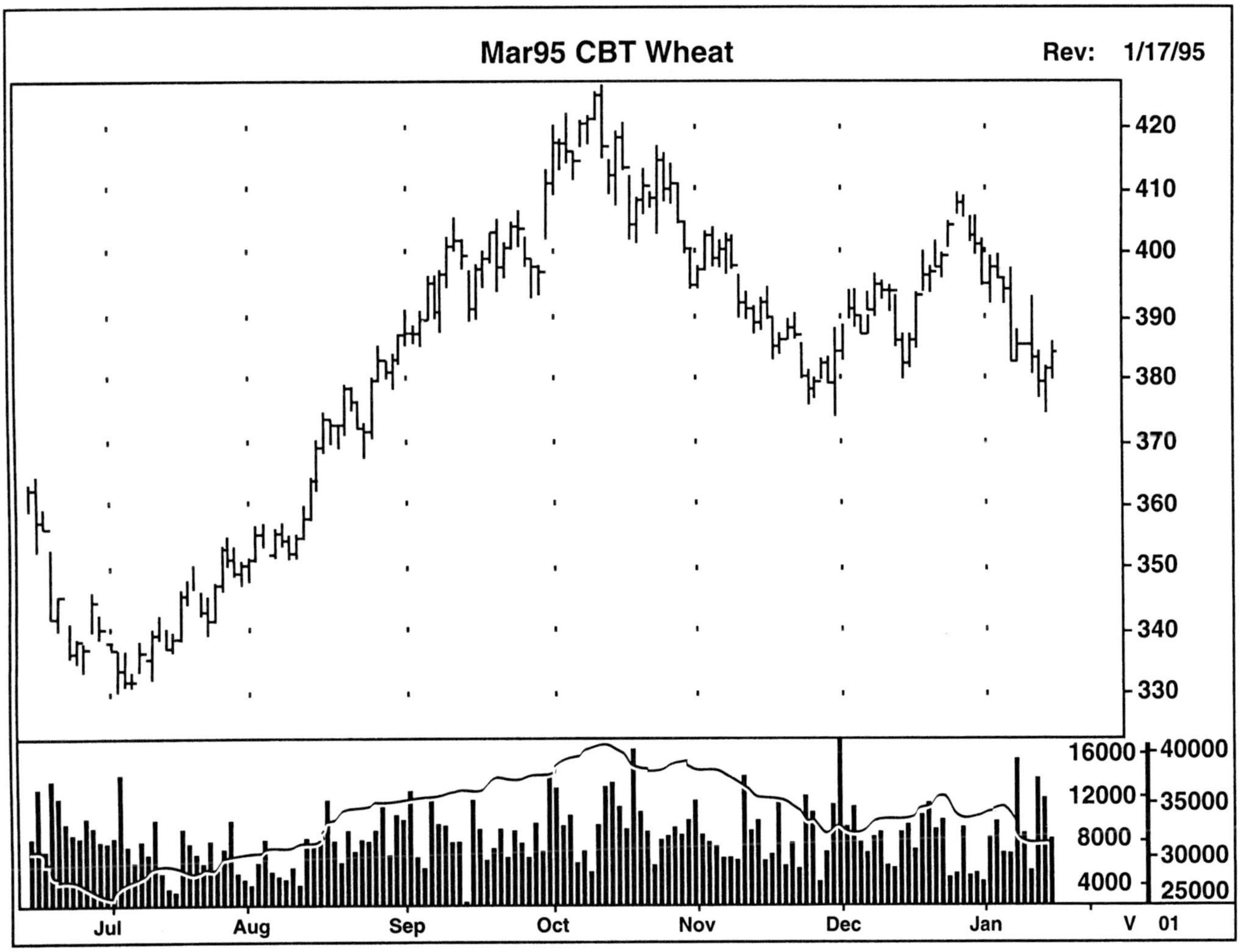

Graph 10 - Bar chart for March 1995 Chicago wheat futures

prices are plotted. Across the bottom of the chart is a daily calendar. Most charts show Monday at the beginning of the trading week. There is no trade on weekends.

On the vertical axis of Graph 10 is a price scale. This shows the prices at which the March 1995 wheat futures contract has traded over an extended time period. Within the chart itself, there are vertical bars and a horizontal dash or "tic" plotted for each day. The vertical bar shows the trading range for the day, ranging from the low to the high. The tic shows the closing or settlement price for the day. This was mentioned earlier but needs emphasis and repeating here. The reader is encouraged to pick up any major newspaper or *Wall Street Journal* and discover that, for each month for each commodity futures being traded, the market shows an opening price, a high price for the day, a low price, and a settlement price. On the bar chart, the opening price is not shown. *The trading range is by the vertical dimension of that plot on the chart ranging from low to high, and the settlement price is determined by the horizontal tic.*

Many producers have price quotes from the futures exchanges coming into their homes or offices. These are updated periodically during the trading day. They also have the capacity to look at bar charts of past price action on the screen or print them out. Many of these services can be maintained for less than $100 per month, and some are in the $30 to $40 per-month range after initial set-up costs are covered. Given the value of information, these fees look nominal in terms of their cost and look small in terms of their relative value to the producer. An alternative approach is to order what is called a hard copy chart service. Typically, the charts are plotted up through each Friday and reach the user via first class mail either Monday or Tuesday. Lacking electronic access to the markets, the user can then employ newspapers, phone messages, or the *Wall Street Journal* to update plots during the week. Then, a new updated chart is available on the following Monday, and the necessary conditions for monitoring the market are present.

In this section, discussion will be restricted to **trend lines, resistance planes, support planes**, and a measure of momentum in the marketplace called the relative strength index (RSI). While this is certainly not an exhaustive list, it is the four types of technical measures that many analysts would agree are of primary importance. Once the use of these techniques is mastered, the reader will then have no difficulty extending usage and coverage into a broader set of technical indicators.

These may be all the tools many producers will need

TREND LINES

A trend line on a bar chart is simply a connecting of two different prices to establish whether the market is in an uptrend or a downtrend. Graph 12 provides an example using the same wheat chart. After trading in a pattern that showed no consistent price movement up or down, this particular contract then traded up for a time before turning in the other direction and moving lower. An uptrend line can be established by connecting the lows at points A and B indicated on the chart. Generally, it is a good idea to keep these two points at least 10 trading days apart and try to avoid drawing steep and short-term lines. That keeps you from being involved in the short-term volatility in these markets. *The market finally closed significantly below the trend line in the area marked C. This is a widely recognized and universally accepted sell signal because it suggests the market is going lower.* For a producer who has been monitoring the fact that prices for his commodity have been moving higher, this is the time to pull the trigger and take action by selling futures to establish short hedges or by buying a put option.

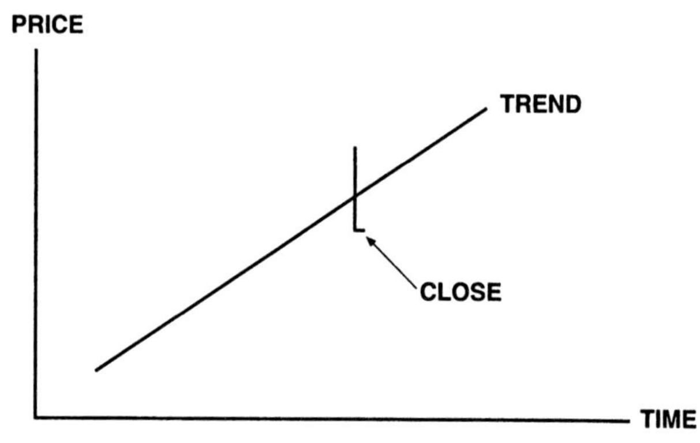

Graph 11 - A close below an up trend line is a widely recognized "sell signal"

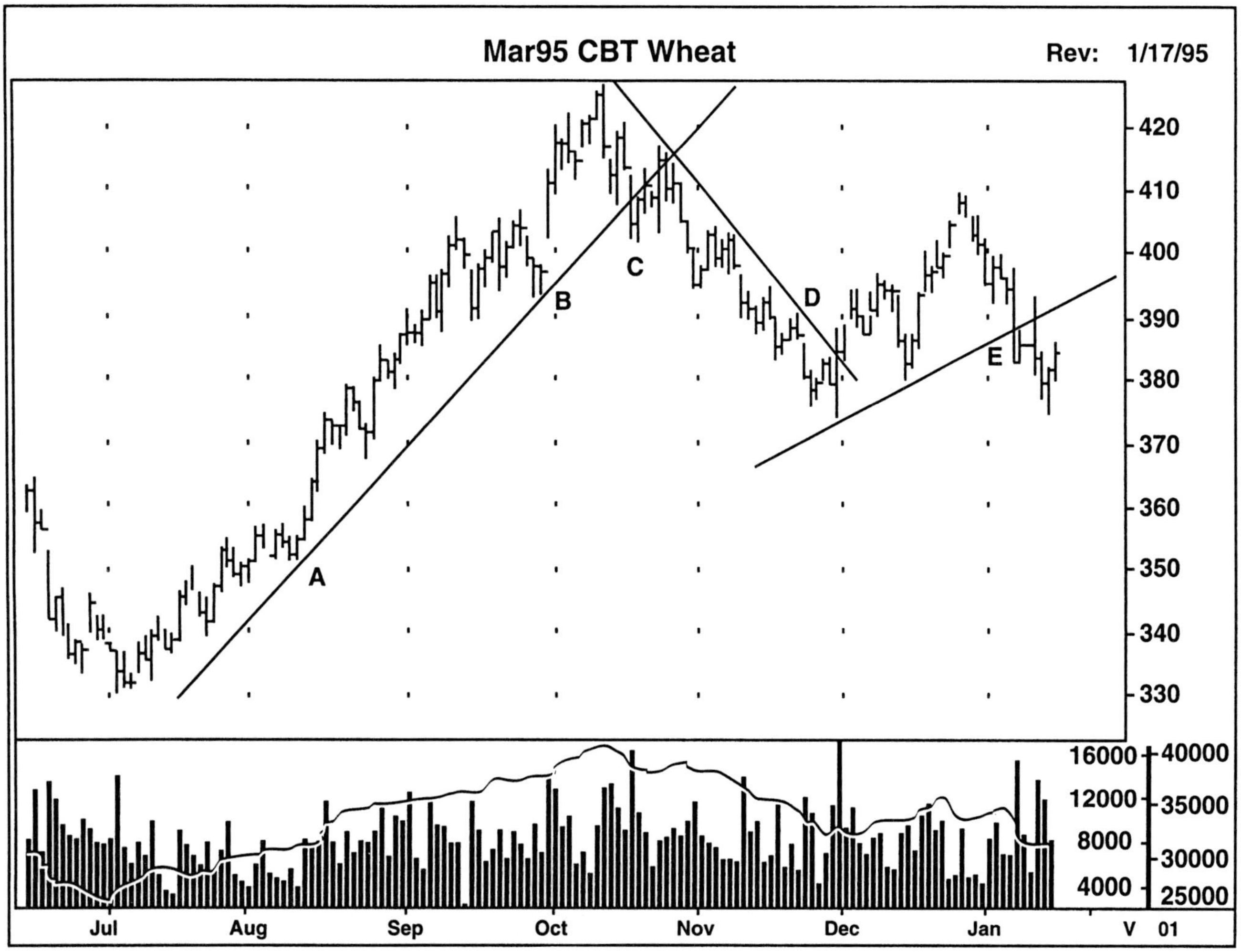

Graph 12 - Demonstration of trend lines and a sell signal at C, a buy signal at D, and another sell at E

Buy and sell signals

It is useful to step back and reflect on what has happened when the market closes below the trend line. Prices had been moving higher, suggesting an underlying strengthening of the fundamental supply-demand picture. Either there is evidence that the supply of this particular commodity is going to be less than had earlier been expected or that demand is going to be better. It could be some combination of both, of course. But eventually, that impetus to the market loses its momentum. The fundamental situation starts to clarify, and there is increased recognition that demand is not going to continue to increase and that the prospects for reduced supply are not going to continue unabated. *At that point, the market lacks a will to trade still higher on the underlying information, and the result is a close below the trend line. That close below the trend line signals a switch in consensus in the marketplace.* Traders who have been waiting for some evidence that the uptrend has run its course will now sell. Traders who were long in the market as speculators or as long hedgers during the uptrend might now also sell to take profits. What we have, then, is a self-fulfilling prophecy to an extent. All the people who are charting the performance in this marketplace, and are not trading strictly on a whim, are going to be inclined to sell the market. It is not surprising then to see the market trade lower in the days following this widely recognized sell signal.

Selective hedging

Many modern producers like to follow what is called a **selective approach** to hedging or forward pricing their commodity. This is one of the advantages of the futures market, for example, as compared to a cash contract. Once you have established price in a cash contract, you cannot then take off that price protection when you decide the market is ready to trend higher in your favor. But in futures you can do just that. The sell signal at C on the chart in Graph 12 is followed some time later by a buy signal at D as a less pronounced downtrend line is penetrated. Note that the downtrend line is drawn across two daily highs that are at least 10 trading days apart. Later still, the market closed below a second uptrend line and appeared to be heading lower. The "selective hedge" would have been replaced near the area marked with an E.

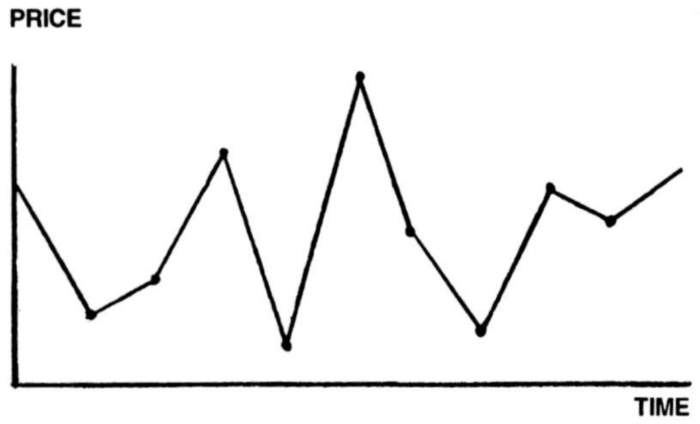

Graph 13 - A selective hedger has the hedge in place only when price is declining

Start simple

What we have, then, is the possibility of the best of all worlds. The producer is not forward priced when the market is trending up, which is to his favor. He has price protection in place when the market is trending down, which is also to his favor. Then, he removes that price protection when the market bottoms so that he can benefit from the developing and emerging uptrend. *If the hedge is placed and held, which is the more simplistic approach, the producer establishes a specific price as has been indicated in earlier chapters. That is the price that he is going to get subject to basis performance. The selective hedge approach opens up the possibility of having protection when it is needed and not having protection, when it is not needed.* He can thus benefit from favorable moves in price.

This is not where the beginning producer should start! We have to keep in mind that it is something of an art to decide exactly which trend line on the chart to draw and use at any particular point in time. We need to keep in mind also that not all markets show significant sustained trends up and then down as is suggested in Graph 12. For the beginner, if price protection is needed, perhaps the best approach is to get the price established and leave it alone until the cash commodity is sold and the futures position is bought back.

The trend line is arguably the single most powerful technical tool that can be applied to the bar chart. It signals when the consensus on the underlying fundamental picture has changed directions and thus gives specific indication to the user as to when price protection should be established. The benefits from monitoring trend lines on a bar chart should be apparent. Once a clearly defined uptrend is established, it is a matter of waiting and watching and not taking action until the market closes below the trend line. There are various types of sell orders that can be employed on most of the commodities to ensure that you will not be short in the market unless the close is, in fact, below the trend line. Any good hedge broker can provide literature on the types of orders. The reference by Purcell at the end of this chapter has coverage on the types of orders and demonstrates their use. *The main need is for the producer to fully understand what he wants to do, make sure that it is communicated clearly to his banker and then to the broker, and then discuss the possibilities in terms of how entry into the market might be completed.* However you get there, the benefits of being short in the market after it has trended upward then turned downward are obvious, and the end result of that type of program is likely to be something the producer likes. Trend lines work.

RESISTANCE PLANE

A second important indicator on the bar chart, especially to the producer who is interested in forward pricing, is what is called a resistance plane. This is simply a horizontal plane sketched across some past daily high in the market. Graph 14 shows an example. Obviously, there can be a number of past daily highs as you look back across several months of trading activity. Therefore, there can be a number of resistance planes. *The resistance plane that is most important is the one at the life-of-contract high.* That is the highest price that a particular futures contract has traded, and it is shown and updated in the electronic systems and in the *Wall Street Journal.* Some larger commercial newspapers also show the contract high.

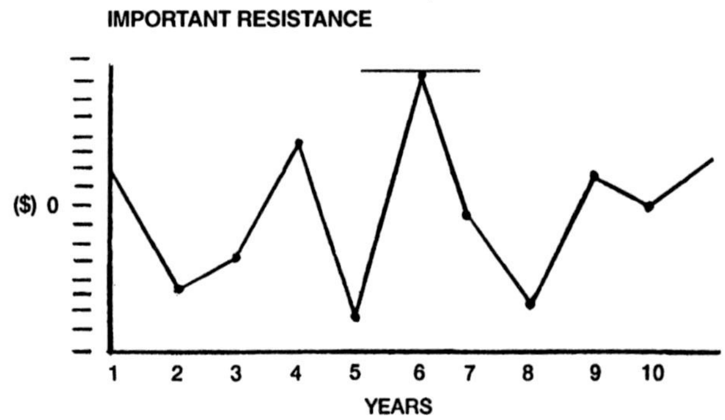

Graph 14 - The resistance plane at the high is the most important

Let's focus for a moment on what happens if the market approaches the life-of-contract high near $3.64 shown in Graph 15. Some time in the past, during late September here, the market has traded up to that specific level. Then, something happens that stopped the interpretations of supply and demand balance from pushing prices still higher. The market turned down. After the topping formation has occurred, there will often be a close below an uptrend line as we had suggested in the earlier section on uptrend lines. Whatever the case, the market turns down and trades down.

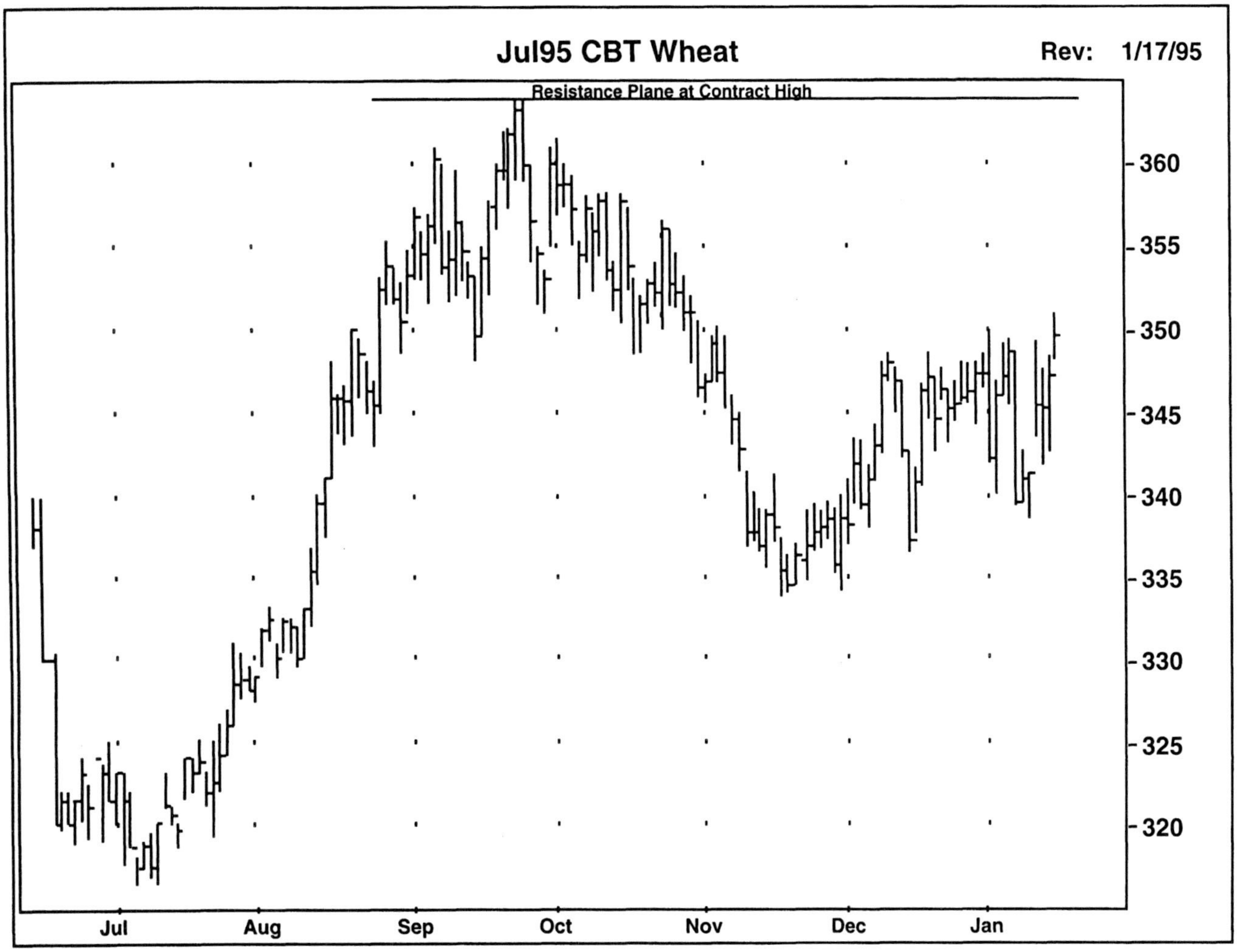

Graph 15 - Resistance plane at contract high for July '95 wheat

Then, at some later time period, the impressions of the supply-demand situation start to improve again. This occurred in November for the July 1995 wheat futures. The cause may be deteriorating weather in terms of crop prospects in the grains and oilseeds, or it may be a *Cattle On Feed* report that shows placements smaller than expected. All of these things suggest future supplies will be diminished, and prices start to climb higher. As a price approaches the resistance plane across the old life-of-contract high, the market faces a very important test. *Will it trade higher still, make new prices, and move into new price ground? Or will traders make the old adage that the market has a hard time going where it has never been before work again?*

A place to sell or forward price

Without any question, when that price is at a relatively favorable level for the wheat producer, the soybean producer, the swine producer, etc., placing sell orders to establish short hedges just below the old contract high is an excellent strategy. In July 1995 wheat, futures prices near $3.60 translate into profitable forward prices for wheat producers. If you step back and think for a moment, you will recognize there is a micro-macro issue as well. Many other producers will be watching the same chart pattern, see the same contract high, and see the same resistance plane.

There is a self-fulfilling tendency at work

They will tend to put sell orders right at or just under that resistance plane, expecting the market to turn lower from there and wanting to get prices established at or near the highs for the year. It is often a good idea to back off $.02 per bushel in corn and maybe $.03 to $.04 in wheat and in soybeans to help make sure that your sell order gets filled. Keep in mind that some sort of distribution of sell orders is sitting there near the old high waiting on the market to trade up there again. Those orders can be placed on a good-til-canceled (GTC) basis, and they are just waiting for the market to come to it. If that creates lots of selling pressure, the market may never equal its old high. *An order placed a few cents under the old contract high is more likely to get filled than an order placed right at or just above the old contract high.* The producer may thus protect himself against the "macro level" impact, the tendency for many producers to have sell orders near the old price high.

Self-fulfilling prophecy

Again, to some extent, there is a self-fulfilling prophecy at work here. Many chart analysts look at the situation and conclude that the underlying supply-demand situation has not improved enough to justify still higher prices. Thus, they expect the market to turn back as it tests the old high and form what is sometimes called a double top in the market as price fails along that resistance plane. Again, unless there is some strong underlying reason for the market to push on to higher prices, it is likely to turn down from here, and the producer as a short hedger will have price protection established at an excellent level. Note that on this wheat chart, an intermediate resistance plane across the highs just above $3.50 will need to be overcome before the market can challenge the highs near $3.64. Producers who, for financial, debt, or other reasons, need to ensure profits, might sell near $3.50 and then sell to price a much larger percentage of the crop if the market does approach $3.64. The resistance planes have a good chance of guiding you to forward prices near the highs of the year!

USING RESISTANCE PLANES	=	TOP PRICES

Pricing on a rally to the contract high resistance plane will tie down good prices

SUPPORT PLANE

The mirror image of the resistance plane discussed in the previous section is the support plane. Drawn across an old daily low, the support plane across the life-of-contract low is especially important. The reasoning is very similar to that presented with regard to the resistance plane but somewhat in reverse. Unless the supply-demand situation has deteriorated badly since the previous low was established, the market is likely to run into buying support at or near that old low, and trade higher again. This is the pattern that is shown in Graph 16 where the market challenged the old life-of-contract low, that very important support plane, and then traded higher again. In October, world demand for soybean oil started to improve. Note the price decline in October stopped just above the old low near $.23.

A place to buy

The life-of-contract low support plane is very important to the user of corn, soybeans, etc. *This an excellent place to place long hedges and establishes the cost of the input of raw material at favorable levels.* Anytime you can get the corn bought, for example, on a long hedge basis down in the $2.15 to $2.25 area, it is such an excellent opportunity that the users should just close their eyes and do it! In fact, they don't have to close their eyes if the old contract low is, for example, at $2.17, and they have something to buy against. This is very important and tends to help bring the needed degree of discipline to the trading activities of the long hedger. In Graph 16, any user of soybean oil had an excellent chance to place long hedges. Prices--costs to the user--were above $.28 in January of 1995.

There is another type of participant in the market that is interested in the support plane across the life-of-contract low. Let's take the situation of a producer who has short hedges in place because he has seen a close above a trend line or he has priced it on a rally against an old resistance plane, and the market then trades substantially lower. As price approaches the old life-of-contract low, the producer might reasonably take a look at the available information and decide there is no obvious reason for still lower prices to be recorded. What he might do then is place a buy order just above the old support plane and look to remove or lift his short hedges and take the profits because he anticipates the market will turn and trend higher, at least for some time period. *This is another approach to the selective hedging idea that was discussed in the section on trend lines.* So what we have is a person holding short hedges in the market who is looking to buy and take profits. We have the speculator who has sold the market and looking to take speculative profits near the support plane, and we have the potential long hedger who is looking to buy near the support plane. Again, the self-fulfilling prophecy issue becomes apparent. *Unless the supply-demand situation has, in fact, turned significantly toward the negative side, we are not likely to see still lower prices recorded by this particular contract. The market is likely to find support and trade higher again.*

A caution here. Using selective hedging sounds too good to be true--and it may be. It is not always easy to know when to place, remove, and replace a hedge. If you left a short hedge near a support plane, you are now a speculator in the cash market. What happens if prices do go down through the support plane--and you can't muster up the courage or discipline to replace the short hedges?

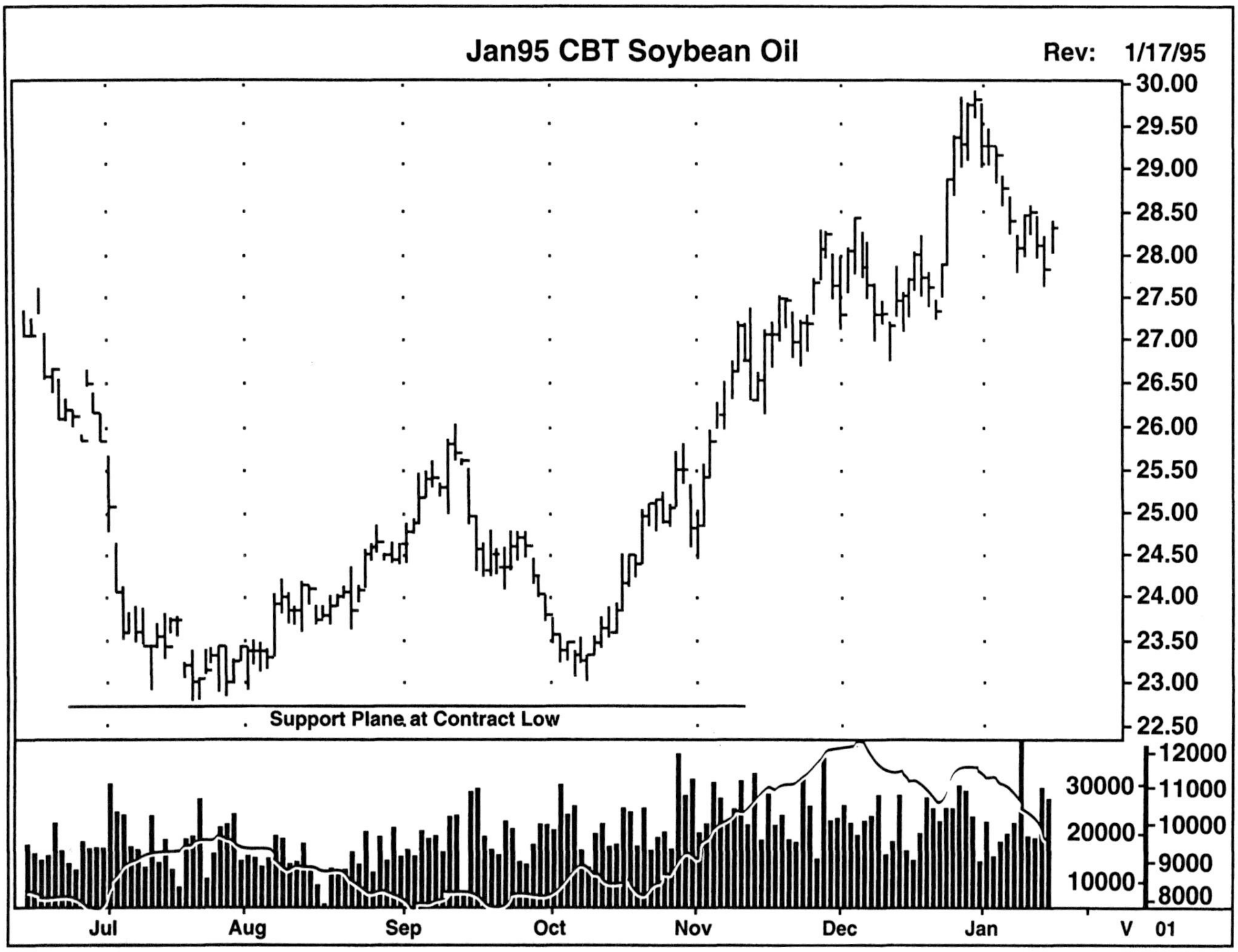

Graph 16 - Support plane at contract low on January 1995 soybean oil

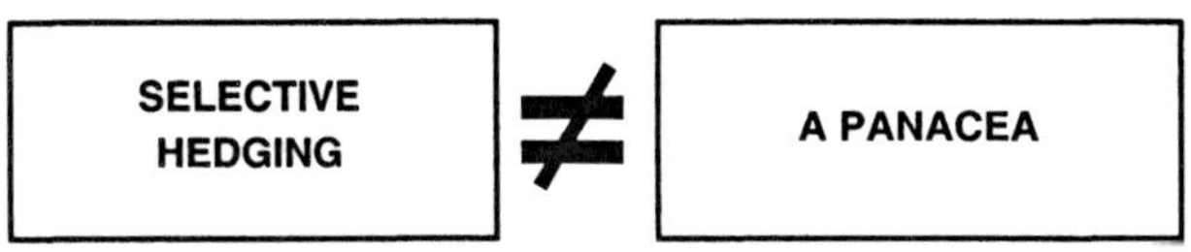

Selective hedging is not the ultimate answer to price risk management

If you put together the three tools to this point, recognizing it is not always easy, you have a trend line to use in those markets where significant trends up and/or down are in fact present. In those markets where no trends are present, resistance planes and support planes can be used to guide timing of hedge decisions. Obviously, a moment's reflection and examination of virtually any bar chart will show that there will be resistance planes and support planes on the chart that are important but don't occur at life-of-contract highs and life-of-contract lows, respectively. Graph 15 showed an example of a past (very recent) high, and therefore a resistance plane, that is below the contract high. But the life-of-contract high was established earlier in the trading life of this contract and was established when the situation and the fundamentally based information was somewhat different. In this instance, the producer who needs price protection might well start to place, at a minimum, sell orders just under that most recent and intermediate level resistance plane to establish short hedges. (This was mentioned above as a possibility for July 1995 wheat.) Then, if the market does in fact trade up toward the old life-of-contract high, the producer answers any margin calls and prepares to forward price a higher percentage of his commodity on a move to the old plane. Thus, it is not just the extremes at the contract high and contract low that allow this approach to chart analysis to be used. The intermediate highs and lows can also be very important.

RELATIVE STRENGTH INDEX

The relative strength index (RSI) is one of the many oscillators that are used by sophisticated commodity traders. The 14-day RSI is shown on many commercial charting services and is available on most of the electronic systems that come into the producer's home or office. What it is designed to do is tell you when the market is approaching or is at a high or when it is approaching or is at a low.

Overbought market

Intuitively, when the market is going up, there will be people who want to buy this particular commodity and get on the bandwagon while there are profits still to be made. This process continues as a market trades higher, and typically open interest, which is the number of contracts outstanding, increases. Eventually, however, almost everyone who is interested in buying corn, cattle, sugar, cotton, etc., has, in fact, bought it. What we have, then, is a relative lack of new buyers or new capital available to enter the market. The market is becoming what is called *overbought*, and it is ready for topping action and ready to turn lower.

Oversold market

On the downside, when the market has been trending lower for a sustained time period, the opposite happens. Virtually everyone who is interested in selling the commodity has, in fact, sold it. The market is *oversold* and is due for bottoming action and due for a turn to higher prices again. Often, the relative strength index will, in fact, signal an overbought condition as the market approaches a resistance plane. It will often signal an oversold condition as the market approaches a support plane. This helps bring discipline to the trader who is trying to decide whether or not, for example, prices are going still higher or whether he should go ahead and get some price protection established at this particular level. *If the market is approaching an old contract high and the market is overbought, then it is time to get the crop or livestock forward priced.*

The overbought or oversold condition of the market is determined by the level of the relative strength index. Graph 17 shows a sketch of what is employed. As the market trades down, the relative strength index tends to come down with it. As it approaches or goes down below an index level of 30, the market is said to be oversold. Conversely, when the market rallies and the relative strength index is pulled higher when it approaches or exceeds an index level of 70, the market is said to be overbought. Thus, the bounds of 30 and 70 on the index tend to indicate the overbought or oversold status of the market. The reader is encouraged to watch the behavior of this index on the charts or on the electronic service that he might be receiving and decide for himself whether or not it is a useful tool.

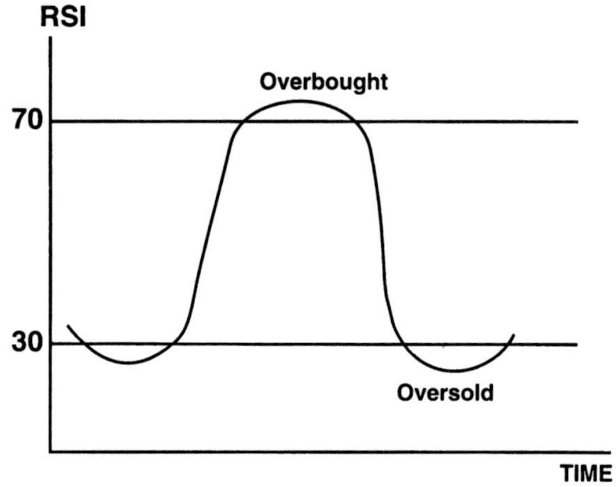

Graph 17 - Relative strength index showing overbought and oversold markets

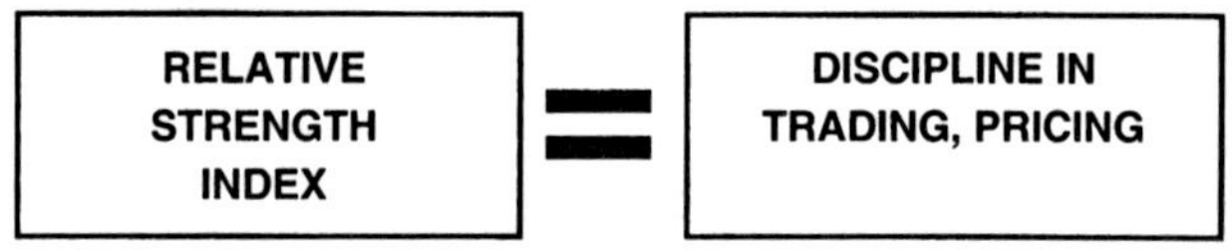

The RSI can help bring discipline to trading by identifying price extremes

Watch weather markets

A caution is in order with regard to the RSI. If we are talking about the corn market, for example, and weather patterns show a prolonged drought, the relative strength index can and will go above a level of 70 and stay there. The notion of having an overbought market probably doesn't have much relevance in a market that is zooming to higher prices because of potential short supplies. The RSI could go to the 80 to 90 range and stay there for a number of days or even for a number of weeks as the weather during July and into early August devastates the potential corn or soybean crop. Thus, it is more nearly regarding normal weather conditions or in markets where there is no dramatic outside disturbance or influence that the relative strength index will be of assistance to the producer.

The complementary nature of the two approaches is starting to be clear

SUMMARY

The fundamental supply-demand balance will determine price. In the short-run, most price moves are due to changes in supply or potential supplies. The producer has to keep up and watch supply-demand reports in the grains and oilseeds, *Cattle On Feed* reports in cattle, quarterly inventory reports in hogs, etc. In the long run, changes in demand can and will change the direction of price trends. It is not difficult to monitor these basic supply-demand measures, especially the supply measures, and get an idea of the likely direction of price moves in the near future. And it is *direction* of movement that is the necessary condition to deciding whether you need price or cost protection.

Technical analysis can help, especially in timing. But the charts can help in confirming likely direction as well. If the chart shows a move up toward a contract high and then a change in direction toward lower prices, this may well help to confirm changes in supply-demand that suggest downward trending prices. Trend lines, resistance planes, support planes, and momentum oscillators, such as the relative strength index, can help establish both direction of price trend and when to act. They bring discipline to the program and are especially helpful with timing of actions.

The need is to put the basics together into a marketing plan that fits the needs of the producer, then follow through. There is huge potential for the producer who develops at least rudimentary skills in fundamental analysis, learns how to read a chart for the most basic technical indicators, and follows through with a sprinkling of discipline. It is much easier than most beginners envision and it can be very rewarding. After all, there is a limit on how much you can sharpen production skills to gain an edge. There is so much room for improvement in marketing that any serious effort is likely to give you an edge.

9

A Changing Marketplace

INTRODUCTION

Much of the discussion in previous chapters has assumed that the individual farmer is operating as an independent producer. Strategies have been discussed and developed to help that farmer handle and manage his or her exposure to price risk. The perception is one of an individual producer who is looking to sell his commodity or his services to a buyer at the best possible price while recognizing that the individual entrepreneur is a price taker. All of that means that it is very important to be aware of supply-demand forces in the marketplace and to have at your disposal some basic marketing tools and use those tools with discipline. We've discussed the need to get your banker involved and to get a broker involved when trades in futures or options on futures are involved. All of this should be compiled into a written marketing plan, and procedures and objectives in that plan might be adjusted and improved over time as expertise is developed. But a very basic point remains.

All of this presumes that the individual producer is in control of his or her exposure to price and to the marketplace. Implicitly, there has been an assumption that most producers are not as well versed in marketing know-how as they are in production know-how. One of the themes and objectives of this entire book is to improve that level of expertise in marketing and marketing strategies.

CHANGES IN THE MARKETS

But the marketplace itself is changing. Change has occurred rapidly across the past 10 to 15 years. The way the livestock, grain, oilseed, vegetable, and other farm commodity sectors are structured and organized has changed. With that change is coming a new demand on the individual producer as a decision maker. Increasingly, it is not enough to just have acceptable levels of proficiency in making marketing decisions and trying to avoid the much-discussed micro-macro trap. Now, it is important that individual producers understand the change that is going on around them and be prepared to adjust and adapt to changing conditions to ensure long-term economic viability.

The markets are changing rapidly

Historically, how various industry sectors in the U.S. economy have been structured and organized has been subject to scrutiny by federal agencies. Going back prior to 1900, national legislation in the U.S. established what is typically called an antitrust policy. In very simple terms, the legislation was designed to protect independent businesses especially relatively small independent entrepreneurs, from unfair competition as they deal with larger firms. This concern extended to the marketplace in which the farmer buys inputs and services and to the marketplace where their product is sold. There is and has been a broad and widely held belief in the U.S. that the individual grain farmer who sells product in hundreds or thousands of bushels may need some protection in dealing with buyers who buy in the millions, occasionally even in the billions, on a nationwide or global scale. When there is a huge imbalance between the size of the buyer and seller, there is concern about whether or not they can bargain in a fair, just, and equitable arena. The large firm, it is often argued, is and financially strong and can generate its own intelligence and its own market information. The small farmer may not be able to do so.

PUBLICLY SUPPORTED MARKET INFORMATION

Over time, the public has tried to improve the relative position of that small farmer by using taxpayers dollars to gather and publish of market information, economic planning information, research at the land-grant universities and within the USDA, etc. Those publicly supported activities were presumed to help even the playing field, so to speak, as the individual producer dealt with large sellers on the input side or large buyers on the output side.

Government agencies provide market information to producers

ANTITRUST POLICY

The antitrust policy has been built over time around a similar notion. There is the idea that when firms get very large, they may be able to exert undue power in the marketplace. In simple terms, market power is defined as the ability or the capacity to influence price or other terms of trade to their benefit because of their large size. Within this environment, requests in the private sector to merge or to acquire competing firms were often scrutinized by either the Justice Department (Justice) or the Federal Trade Commission (FTC). During the

1950's, 1960's, and 1970's, there were innumerable merger and acquisition requests coming from the private sector that were denied by the federal agencies, and occasionally the issue was decided in the courts. During the 1980's, the Reagan administration moved toward deregulation and less involvement by the federal government in the marketplace. The number of merger and acquisition requests that were denied or blocked by Justice or FTC declined sharply in the 1980's compared to earlier decades. Merger guidelines were developed and released by Justice in 1984, and the decade of the 1980's was one of active moves in the private sector toward consolidation and toward the large size that mergers and acquisitions can provide.

Year	Steers & Heifers	Sheep & Lambs	Hogs	Boxed Beef
		Percent		
1978	30	56	34	50
1979	35	64	34	51
1980	36	56	34	53
1981	40	52	33	57
1982	41	44	36	59
1983	47	44	29	60
1984	50	49	35	62
1985	50	51	32	62
1986	55	54	33	68
1987	67	75	37	80
1988	70	77	34	79
1989	70	74	34	79
1990	72	70	40	80

Table 1-The 4-firm concentration ratios for selected classes of livestock slaughter and meat production

A widely used measure of concentration in any particular sector is what is called the CR-4 ratio or the 4-firm concentration ratio. This shows the percentage of business activity in any particular sector that is being handled by the four largest firms. Table I shows change across time and CR-4 measures in some of the livestock and meat commodities as they changed during the 1980's. Much the same thing has occurred in grains, in oilseeds, and in fresh fruits and vegetables. But the change across a fairly short time period is arguably more dramatic in livestock and meats than in other agricultural commodities.

Note that in 1990, the CR-4 ratio is at 80 in boxed beef activity and is at 70 or higher in steer and heifer slaughter and in sheep and lamb slaughter. This suggests that four large firms are doing 80 percent of the business activity in some sectors, and there are indications that the CR-4 ratio is starting to increase in pork and in other sectors as well.

CHANGE IN THE 1980'S

Many of the actions taken in the 1980's were supported at Justice and FTC on grounds of benefits from economies of size. This important economic notion has been introduced early in the book and, without question, it is important. In virtually any type of production or processing activity in the agricultural sector, the large firm can do it cheaper than can the small firm. Graph 1 offers a useful review of this showing estimates of cost of slaughtering cattle in plants that range up to a level of 325 head per hour. Note that costs are still declining at 325 per hour and that the costs of slaughtering on a per-head basis at 300 to 325 head per hour are sharply below the costs of slaughtering in the 50 to 100 head per hour range. Presumably, the benefits of these reduced costs get passed back to cattle producers in the form of better prices and, in the long run, get passed forward to consumers in the form of an increased supply of beef at lower consumer-level prices. There is no doubt that such benefits accrue from economies of size, and they have been widely documented in a number of research efforts.

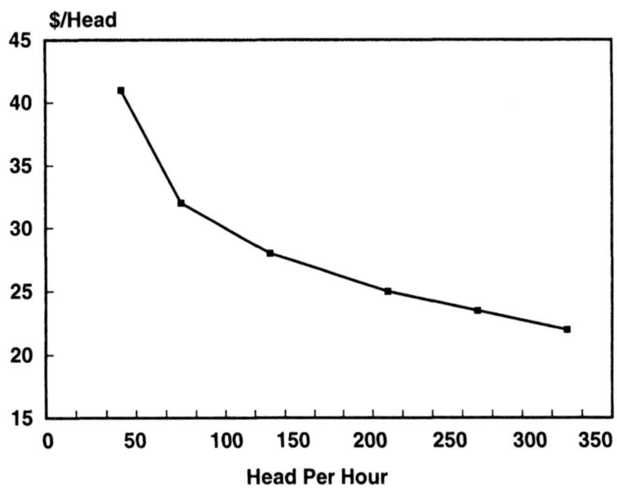

Graph 1 - Estimated costs of cattle slaughter by size of operation

The controversial part of the changing marketplace arises because along with the move toward a few large firms has come significant changes in the way the marketplace is configured and the way the individual producer has to deal with sellers and buyers of their livestock or crop commodities. Perhaps because the large firms are in a position of more nearly expecting and getting change at the producer level, we have seen massive moves toward either vertically integrated activity between the producer and the processor and/or contractual arrangements that bind the producer and processor together in a significant way. Both of these moves mean that a level of control is now available to the processor that was not there historically. It is this very important change in the marketplace that needs to be discussed in more detail in this chapter, and it is this change that the individual producer needs to understand and adapt to.

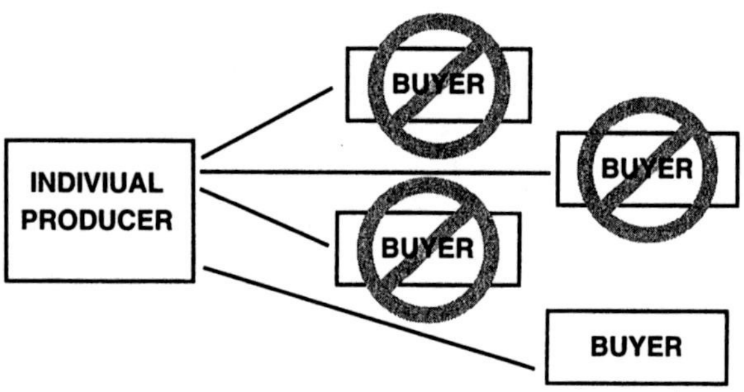

Producers have fewer potential buyers in the changed marketplace

TYPES OF AND REASONS FOR CONTROL

In a changed marketplace, there are two broad types of control that are now becoming commonplace:

- **Vertical integration**
- **Contract production**

Both have significant implications to the marketing decision and marketing strategy of an individual producer.

VERTICAL INTEGRATION

The first and perhaps most extreme approach is what the literature calls *vertical integration.* This terminology is usually employed to refer to circumstances in which two or more of the functions between the producer and the consumer have been brought under ownership control of the same business firm. It doesn't always have to be from the processor back down toward the producer, but that is typically the way it has gone. Processors have gained ownership control over production activity, and they closely coordinate what is produced with what they need at the processing level. *This means that instead of relying on the price mechanism to give signals to a producer to change qualities or change grades or adjust over time and to provide the product in a timely way, forcing coordination through ownership-type control and non-price means.*

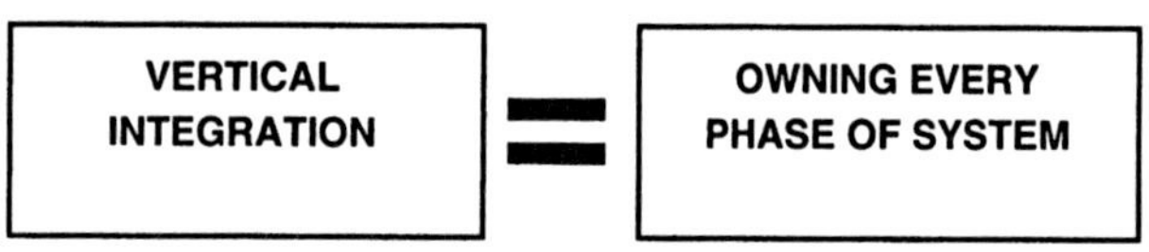

Vertical integration allows one manager to control all the activities

CONTRACT PRODUCTION

A second and broadly employed approach is to use contracts. Processors, for example, contract with producers and specify quality, timing of shipment, perhaps price considerations, etc., to help ensure that what they get is what they want and when they want it. This is a less extreme approach to forcing coordination between the various stages than is vertical integration, but it is a rapidly growing approach that is seen all through the agricultural sector.

QUALITY CONTRACT

The "why" of these changes in the marketplace is basically two-fold in nature. Perhaps the most important rationale for processors is to gain control over the quality of the raw material that goes into the processing facility. Without any question, processors of hogs, tomatoes, cotton, peanuts, or any other agricultural commodity have faced a highly variable flow of raw material in terms of quality. As firm size grew and as processors started to think about how they might be able to flex more of their muscle, one of the things that has happened is that processing firms have started to get more interested in product development. They want to better merchandise their product and start to think about putting their own brand on the product. While branding has been there for a long time in oranges, bananas, and some of the fruits and vegetables, it is relatively new in the meats, especially in fresh meats, and in some fruits and vegetables. Generally, the steak, pork chop, or other fresh meat cut at the retail level does not carry any brand or identification other than that of the food store in which it is sold. This is not true of the poultry sector, which has been vertically integrated for many years. Most of the integrated poultry firms do brand their product. They have moved in that direction as they gained control over decisions on genetics and in other ways brought the quality variability under control.

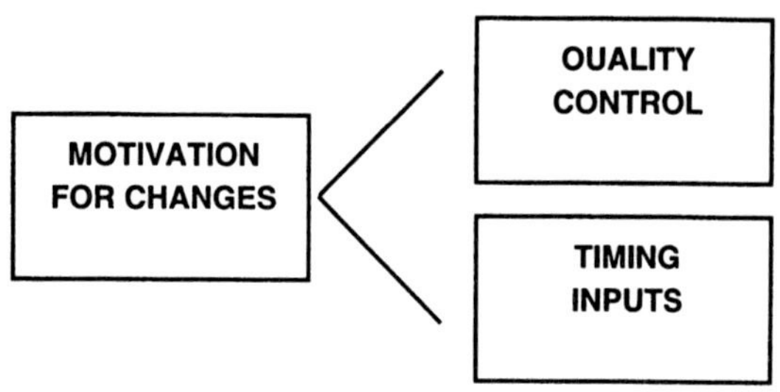

Gaining control over quality and timing of input flows are two important motivations

PRICE SIGNALS

Theoretically, the price mechanism is supposed to be able to do all this. If the processor or retailer needs a different quality, or some change in the product that is being provided by the original producer, a price premium is sent down to ensure that the desired product attributes are offered. A *price discount* gets sent down through the system to ensure that the producer who is offering the wrong product attributes gets the message and changes. A *price premium* is supposed to get sent to producers of the grade or quality of product consumers want. The producers then produce and offer more of that grade or quality.

In fact, the effectiveness of the pricing mechanism in accomplishing all this is modest at best. There have been efforts for decades to try to get a pricing mechanism, for example, that truly rewarded the producer of a high-value product and discounted prices to the producer of a poor product. But those efforts have really never been very effective. Where large processing firms have the capacity to exert influence, they have started to control that quality variability. Moving to vertically integrated structures, or contracting, have been the prevalent ways to get this done.

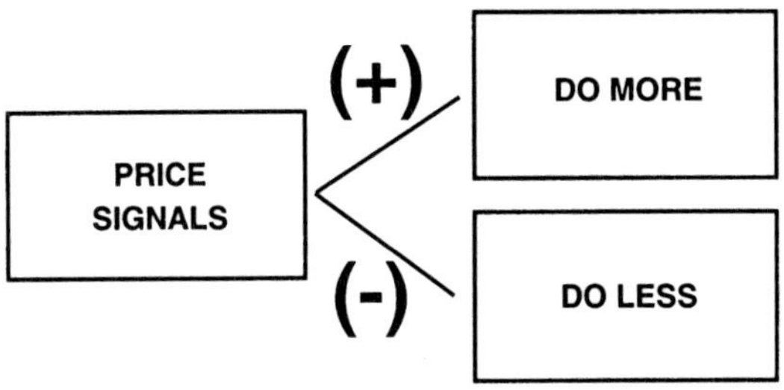

Price signals are supposed to prompt producers to change

QUANTITY CONTROL

A second, and perhaps equally important, motivation for getting control relates to the timing of the flow of materials into the processing facility. In Fig. 1 we showed the cost of slaughtering cattle in a plant that is slaughtering an almost unbelievable 325 head per hour. That involves a complex assembly line type of activity within the plant with a worker or workers at various stations along the line. The process starts with the animal being brought into the plant and ends when closely trimmed boneless retail cuts of beef are placed in a box and loaded in a truck at the back door. It may cost the plant $3,000 per hour in out-of-the-pocket or variable costs to run this line. If 300 head per hour are being slaughtered and the variable cost is $3,000 per hour, then the variable cost of slaughtering is $10 per head. If, because the cattle are not readily available in a timely fashion, the line speed has to be slowed during the day or for one day during the week down toward 200 head per hour, virtually every cost of running that line will still be present. You still have to have a worker or workers at every station, and most of these employees are working under either a union or a company contract where they are guaranteed a 36 to 40-hour work week once they show up on Monday. *It is obviously important, then, that a flow of the right type of animal be there on a timely basis.* If the line speed has to go down to 200 head per hour, if that cost is still $3,000 per hour then the per-head cost of slaughtering increases 50 percent to $15 per head.

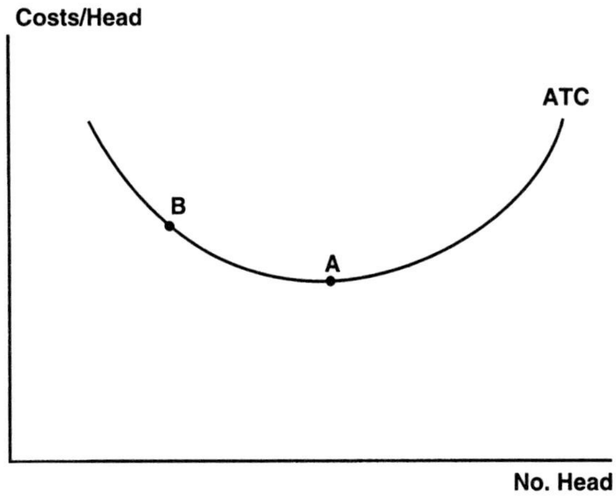

Graph 2 - Example of an average total cost curve

There are obvious reasons why the processing firm would want to control the flow of product. Graph 2 shows a U-shaped average total cost curve to give the idea of what is going on here. If the flow of raw material into the plant can be kept at A, which is near the bottom of that average total cost curve and near the designed capacity of the plant, costs will be substantially lower than if the plant has to operate at B. Obviously, the processing facility needs to try to ensure an orderly flow of raw material into the plant. If the only way they can get this done is to either control the production themselves or to contract with tight specifications on when the product has to be delivered, they will do so. The result is a change in the marketplace compared to what we had seen historically and, as suggested, a substantial change in the role of and in the opportunities facing the individual producer.

Production and Marketing Contracts						Integrated Ownership					Total				
	1955	1960	1970	1980	1990	1955	1960	1970	1980	1990	1955	1960	1970	1980	1990
	Percent					Percent					Percent				
Livestock:															
Broilers	80	91	92	91	92	0	5	7	8	8	80	96	99	99	100
Turkeys	32	46	60	62	65	1	4	12	28	28	33	50	72	90	93
Hatching Eggs	55	60	70	70	70	20	30	30	30	30	75	90	100	100	100
Marketing Eggs	13	21	35	48	43	2	5	20	45	50	15	26	55	93	93
Mfg. Grade Milk		25	25	25	25		2	1	1	1		27	26	26	26
Fluid Grade Milk		95	95	95	95		0	0	0	0		95	95	95	95
Hogs		1	1	2	18		1	1	2	3		2	2	4	21
Fed Cattle		10	18	10	12		7	7	4	4		17	25	14	16
Sheep/Lamb		2	7	7	7		5	12	9	33		7	19	16	40
Field Crops:															
Food Grains		1	2	8	7		1	1	1	1		2	3	9	8
Feed Grains		0	1	7	7		1	1	1	1		1	2	8	8
Cotton		5	11	17	12		3	1	1	1		8	12	18	13
Specialty Crops:															
Pro. Vegetables		60	85	83	83		8	10	15	15		68	95	98	98
Fresh Vegetables		20	21	18	25		20	30	35	40		40	51	53	65
Potatoes		35	45	60	55		20	25	35	40		55	70	95	95
Citrus		40	55	65	65		20	30	35	35		60	85	100	100
Other Fruit		15	20	35	40		15	20	25	25		30	40	60	65

Source: Patrick M. O'Brien, ERS, USDA, NPA Conference on the Future of the U.S. Food System, April 1-2, 1993, Kansas City, Missouri.

Table 2-Alternative marketing mechanisms, contracting and integration

Table 2 shows what has occurred since 1955, with activity in various types of vertically integrated structures or contracts shown for 1960 through 1990 at 10-year intervals. If you add across the line, notice that broilers are essentially 100 percent produced either under integrated ownership or (and this is where most of the volume is) under contract. Hatching eggs also show 100 percent. These are the sectors that historically, even as far back as 1960 to 1970, had shown substantial amounts of either vertically integrated activity or contractual activity. Most of the changes in recent years have come in areas like hogs and in some of the specialty crops shown near the bottom of the table. Notice that the percent of hogs produced under contract was at 2 percent in 1980 and was up to 18 percent in 1990. In the mid-1990's, that number would be substantially higher as would the 3 percent number for hogs produced under integrated ownership. Note that if we combine contract and integrated ownership production in the hog sector, the number jumps from 4 percent in 1980 up to 21 percent in 1990. In 1995, that number is in the 30 to 35-percent range.

It is interesting to note that in the livestock commodities, much of the production is done under the contract alternative. In specialty crops such as vegetables, citrus, and some others, a substantial proportion of the production is done under integrated ownership. Notice that the citrus production near the bottom of the table shows 100 percent when the contracts and integrated ownership are combined, but a much higher proportion of this is in the form of integrated ownership than was the case for, say, hogs or for poultry. In processing vegetables and fresh vegetables the total is 98 percent, but 83 percent in 1990 was under contract and only 15 percent in integrated ownership. The patterns differ across the different commodities, but the motivation is the same. The processing firm badly needs to get a consistent flow of product into their facility when they need it and of the quality they need.

It would appear these trends are going to continue. The interesting question in terms of the focus of this book is what does all this mean to the individual farmer? What should the individual producer do? How does it change all of the discussion that has been completed in earlier chapters about marketing and marketing decisions? In particular, what should the individual producers think about doing in this changed environment in order to secure their futures and make sure they are economically viable?

Producers will need to decide how they will react to a changed marketplace

Some producers will like contracting with buyers

INFLUENCES ON THE MARKETING DECISION

The precise type of influence on the important marketing decision depends on the type of program that is facing individual producers. If they are producing under contract for a processing firm that makes most of the market-related decisions and handles any exposure to price risk, then individual producers essentially have no marketing decision to make. What they are offering is their management expertise and their ability to produce efficiently, to meet pre-set quality specifications, and make sure that the operation is efficient and low cost in nature. The critic would suggest that this is moving the individual producer away from an entrepreneurial role toward one that more nearly looks like a "hired hand." This has been one observation regarding the long integrated poultry sector, where the poultry producer typically makes the investment in the building and then has a contract relationship with the integrated processor that provides premiums for superior management performance and penalties when performance is not so good. Without debating whether this is good or bad for the individual producer, it is clear that it switches the emphasis away from concern about being an effective market analyst to one of being at the cutting edge in terms of production technology and production know-how.

Not all contractual relationships, of course, protect the individual producer from exposure to price risk and to the marketplace. Many approaches to contracting that are now starting to emerge require the producer to be extremely well informed about marketing.

CONTRACTING APPROACHES

One example would be a contract that specifies quality characteristics and provides for premiums or discounts based on quality. The primary focus of attention is scheduling the product into the processing facility. For our purposes, we might just call this a **time contract**. No price or provisions for price is included in the contract, and this means that *the individual producer is totally open to price risk in the marketplace.* All of the hedging and option strategies that were developed in earlier chapters would be appropriate and would apply here.

Time only

For example, if a producer has a contract with a large swine processor that lays out a premium and discount schedule for quality of hogs and specifies a five-day time period within which the processor can ask for the hogs, the producer might still be totally exposed to price risk. This is no different than operating in the open marketplace as a entrepreneur except the buyer of the hogs has been established and the shipping time period has been determined. This might be quite attractive to a progressive producer who wants to make sure that he negotiates a premium for his hogs. He knows they have the type of quality that the processor wants and would still prefer to function as an entrepreneur. In other words, he or she brings ability as a market analyst to the timing of the pricing decision, and this is very much consistent with the theme through earlier chapters in this book.

Basis contract

Another approach to contracting that is growing in usage is what is called a **basis contract**. This notion has been introduced in earlier chapters. The time period for delivery of hogs or grain, or whatever the case, may be determined, and a cash-futures basis negotiated, in the contract. The buyer of corn in central Illinois, for example, might be wanting to buy at $.15 per bushel under a particular futures. The producer might be able to negotiate a slightly better deal than that and get $.12 under a particular futures. The important thing that has been specified in the contract, then, is the relationships to a specific futures contract. In all likelihood, time at which the grain will be delivered and a set of premiums and discounts for quality, percentage of foreign matter, etc., will also be included. *The producer is still totally exposed to price risk.* All of the fundamental and technical approaches to analysis that have been discussed in prior chapters need to be brought to bear here.

It is good to have what appears to be, based on past records, a favorable basis tied down in the contract, but having a favorable basis tied down doesn't do much good if the corn market goes to $2.00, well below cost of production. Thus, it is a matter of monitoring the situation and taking advantage of that contract provision that allows you to call on a particular day and indicate when price is to be settled. You would then get $.13 below the settlement price for the appropriate futures contract for that particular day. *It is very important that producers who enter into basis contracts in livestock, grains, and oilseeds understand that they have no price protection and that the only thing they have accomplished is possibly the negotiation of a cash-futures basis that is relatively favorable.* This eliminates any of the basis risk we talked about in earlier chapters. But *price risk* is still present!

A BASIS CONTRACT = **FIXES THE BASIS ONLY**

A basis contract ties down basis only, not price

Formula price

Yet another approach to contracting agricultural products is what is called a ***formula price contract***. Also mentioned in earlier chapters, this has been subject to much discussion across the years, and in very recent years has been bitterly attacked by some fed cattle producers as incorrect and wrong. What happens is that a particular livestock, vegetable, or grain producer might agree to deliver the product during a certain time period at a price that either shows some premium or some discount to a widely recognized cash market price. Note the difference here: Instead of tying to a futures contract as in the basis contract, this type of contract may be tied to a cash market price. For example, the producer of hogs in Ohio may negotiate a contract with a hog packer/processor to sell hogs, of some specified quality, at $1.00 per hundredweight above the Southern Minnesota/Iowa price. This latter is a widely reported and widely watched price series for hogs, reported daily by the USDA. The Ohio producer's price is essentially tied to that reported price.

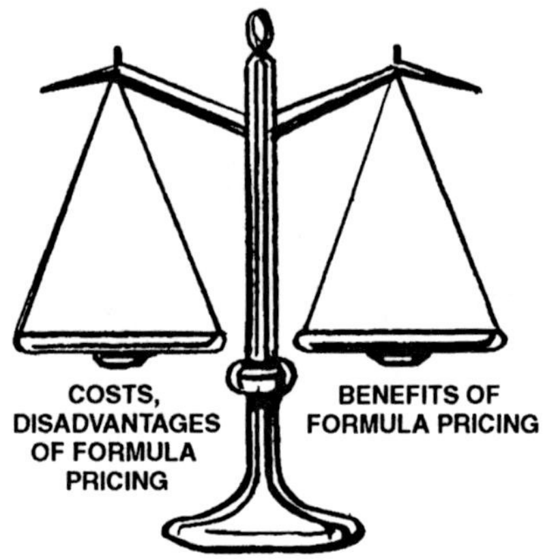

It is important to weigh the good and bad of formula pricing

The criticism of this approach comes because some producers feel that the producers involved in formula pricing are giving up their right to influence the price discovery process and giving up what some producers would argue is their responsibility to keep the pricing process more open. But times have changed, and many producers find this approach of negotiating prices relative to some widely observed market indicator around the country and effective one.

Keep in mind that this type of formula pricing again leaves the producer totally exposed to price risk. The relationship to some widely observed and widely watched cash market has been established, but there is nothing to keep all of those prices from going down and hurting the producer's position. Thus, all of the strategies involved in hedging, put options, etc., that have been discussed in some detail are appropriate. It may well be that the producer has to get involved in a formula price arrangement with the packer to protect his access to that packer as a buyer. *The burden of responsibility in terms of protecting himself against price risk is still present and very much an important issue for the producer.*

HOW TO BE COMPETITIVE AND SURVIVE

Questions that are always asked in this changed marketplace include: What is the role of the individual entrepreneur? What does the producer do? How does he or she survive? What is there in the changing marketplace to ensure there will be a place for the relatively small independent producer?

THE SIZE ISSUE

The answer to all of these questions is not a simple one. Being small and being viable is, in some instances, going to be difficult just because of the economies of size. That is not to say that an Iowa farmer who has cash corn and soybeans and 50 sows on his farm cannot stay in the hog business. There will be some disadvantages. Fifty sows are not enough to offer a truckload lot of hogs, and with an operation this small, chances are that less attention will be paid to genetics and staying on the cutting edge than would be the case if he had 1,000 sows. Still, the small operator may be able to spread fixed costs, including family labor, across the various enterprises and compete quite well. But realism dictates that we recognize that the small operator will, in general, face a tougher challenge in terms of surviving and competing over time than will the somewhat larger operator.

WORKING WITH PROCESSORS

There is always some debate on why the modern swine processing facilities, for example, are moving into production and why some processors of fruits and vegetables have moved aggressively into owning their own facilities over time. It may be that production in these particular areas is just simply quite profitable over time if good management and good technology can be brought to bear. A very important set of reasons, as discussed above, is the need to get the quality variation under control and the need to be able to schedule the timing of raw materials into the plant to keep these new and large plants operating at or near capacity. *If the individual producer will learn what quality the processors want and try to work with them in terms of when they need it, then it may well be that the processor will let the producer make the capital investment in expanded production over time and just stay in the processing business as compared to getting into the production business.*

Think about meeting the needs of the buyer

POOLING GENETICS

There are points here worth additional explanation. It is not just size that will make the producer attractive to a processor. There has been much talk about the need to put together things in truckload lots so that the processor will be interested. For small producers, however, it is not just a matter of pooling their product with other small producers. You have to be careful with pooling arrangements. It doesn't matter whether it is in hogs, vegetables, or something else, you are probably going to get more variation across producers than would be inherent to one large producer. Processors know this. *In addition to pooling volume of product and pooling trucking and other types of functions that several small producers can cooperate on, it may also be a matter of pooling the genetics.* This is true whether it is peanuts, fed cattle, or hogs. The primary source of quality variation at the final end of the processing is from variation in the raw material. Producers may have to recognize that they will need to pool and operate and work together in some ways that they have not necessarily been involved in before.

Obviously, it is going to be important to do a good job of marketing. There are many alternative arrangements that involve meeting the processor's need such as premiums and discounts on negotiated prices that are tied to quality and the timing of the raw material flows. That still leaves the producer, all too often, totally exposed to market risk. Most prefer it this way. This means they are truly entrepreneurs as compared to being the "hired hand" that some people feel the contract producer has become. If you are going to meet the needs of a changed marketplace, one of those needs is doing a still better job of marketing and managing your exposure to price risk. *That entrepreneurial function is still very much in place and is still very important.*

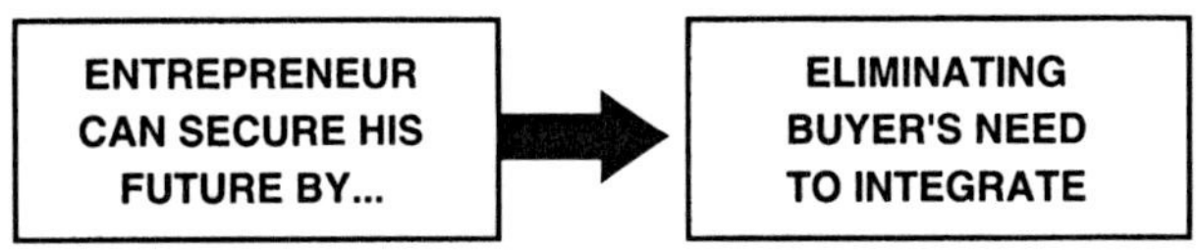

Some changed attitudes and a willingness to work together are needed

DIVERSIFYING HELPS RISK

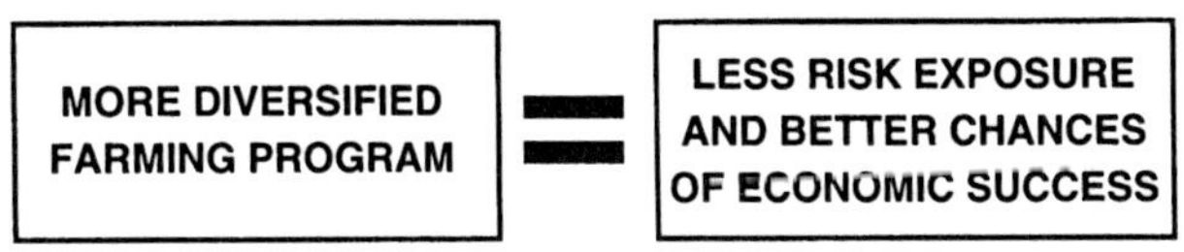

Diversification is survival insurance

Another approach to ensuring your long-term survival as an entrepreneur is to diversify. This age-old advice suggests you don't put all your eggs in one basket. You diversify what you produce and what you do. Part of that diversification might be to get involved in contract production with an available buying processor, yet in other commodity areas stay in the open marketplace and function totally as an entrepreneur. This may mean, for example, that you are contract producing hogs for a processor and producing a vegetable crop that you are marketing directly to a local regional food chain because you can get a better price that way. All of these will need examination. One of the criteria that should be employed in determining when you do and don't contract is your exposure to price risk. It may be important, for example, to tie down a buyer and make sure you have a "home" for a vegetable crop that could be ruined if not sold within a particular week. It may be less important to contract a corn crop which you could put in on-farm storage bins and try to use your knowledge as a marketing analyst to determine when you sell the crop. Thus, diversification can mean different things for different crops and different things for different producers, and contract production is one way you ought to think about diversifying.

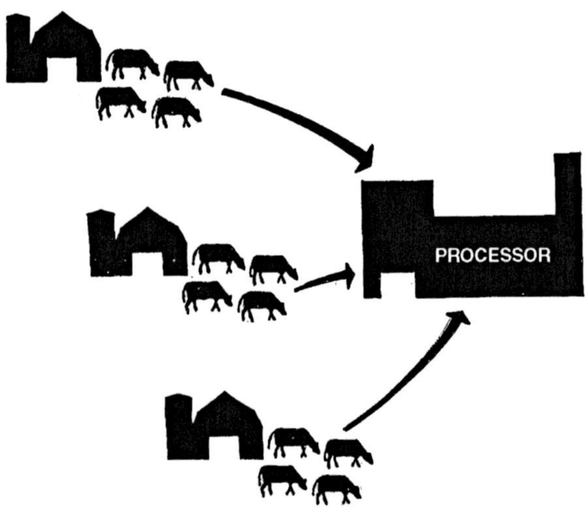

Industrialization of agriculture means changes for the producer

SUMMARY

We are seeing what has come to be called "industrialization" of agriculture. Many farmers don't like what they are seeing. The marketplace is changing. The trends that we are seeing toward vertical integration and toward contractual arrangements are not likely to be reversed in the short term. We are not likely to see a sharp reversal of antitrust policy at the federal level, which allowed substantial consolidation and moves toward large firm sizes at the processing level during the 1980's. In other words, the ball game that is out there in the mid-1990's is likely to be roughly the ball game that will be there in the years 2000 and 2005. These trends that we are seeing will have carried us further toward a changed marketplace by 2005.

The need for quality control at the processing level will not go away. The need to schedule the flow of raw materials into large and efficient processing facilities will not go away--it will accentuate. These trends are permanent, these trends will persist, and the individual producer needs to step back and reflect and not get caught in fighting inevitability.

Producers need to understand the issues, however. They need to understand we are now in a global market, that what is happening in other countries is important. And producers need to recognize that there is benefit to coordinating activities. They need to get past the long-standing, often adversarial, relationships between the producer and the processor (seller and buyer) that can be found all through agriculture. We need to recognize that coordination can make all of agriculture more profitable. That is why the poultry firm integrated initially, that is why we are seeing integration and contractual arrangements come into the pork sector. That is why we have seen it happen in fresh fruits and vegetables, and that is why we are starting to see it happen increasingly in the traditional grains and oilseed. *There are benefits to coordinating.* It is important that every function in the assembly line between producer and consumer be effectively coordinated.

Any entrepreneur operating at any one of those stages of economic activity fights the trend is going to be in trouble. In some instances you have to go along to get along. Fighting a processor is likely to mean you end up without access to a very important part of your marketplace. It is going to take some changes in attitudes, and it is going to take some compromise. But that is always the art of good politics, and it is effective in this changing marketplace as well.

Through all of this, the producer needs to be a good marketer. We have seen historically much more emphasis on production technology and production know-how than on marketing technology and marketing know-how. That historical emphasis will stand the producer in good stead now as we move toward contractual programs with processors in which there will be rewards and benefits accruing to production management expertise. But often, there will still be exposure to price risk. In markets that are more volatile than they were 10 to 20 years ago, and in markets that are increasingly exposed to a world market and all of the vagaries that come with it, it is incumbent on the producer to be a good marketer. *Arguably there is far more potential for improvement in the financial and economic viability of the individual farm through improved marketing than through improvements in production know-how and expertise. This suggestion was offered earlier, and it is still valid in a changing marketplace.*

10

Final Thoughts and a Bit of Motivation

INTRODUCTION

The marketplace is changing. Chapter 9 talks about how it is changing in the United States for the American farmer. It appears that we will have to get accustomed to operating with different procedures and with different types of exposure to the marketplace. The producer is still one of many relatively small producers and is still, in virtually all of the important agricultural commodities, a price taker. About the only way to avoid that position is to get into a niche market or into a marketing effort where some program is involved to differentiate the product. We will see more of this in the major commercial commodities. It is not going to be unusual in the future, for example, to see identity of the producer of corn or soybeans maintained through marketing because that crop has been produced to meet a certain protein level requirement or some other inherent characteristic. *But the vast majority of our production into the world market is still going to be produced in an environment where the individual producer is definitely a price taker, and the marketplace in the U.S. is changing for those entrepreneurs.*

Governmental policy changes have altered the marketplace

A WORLD MARKET

Increasingly, it is not just changes in the U.S. that will impact on the American farmer. We are increasingly exposed to the world market and all the vagaries of weather and political decisions that come from that exposure. It hasn't always been this way. In the 1960's, for example, we were operating the in the U.S. under an entirely different type of marketplace. Government programs controlled the level of production by assessing acreage quotas, and price received by the farmer was guaranteed by federal government programs at some percent of parity. The parity price had been legislatively determined and tied back to the rather ancient 1910 to 1914 period when there was a perception that farmers and non-farmers were faring about equally well in the U.S. economy. Under those conditions, there wasn't much reason to learn about markets and marketing. There wasn't much to be gained by the typical American farmer because there was no effective marketing decision to make. Price risk was negligible, and the exposure to a very volatile selling price that has occurred in later years was simply not present.

It was policy change both in the U.S. and in other countries that started the trend toward the marketplace that we see today. It was policy decisions, then, that brought on the need for the American farmer to develop expertise as a marketer. It is interesting in this final chapter to reflect on what those policy decisions were, why they were made, and what it means.

POLICY DECISIONS: IMPACT IN THE MARKETPLACE

In 1973, a policy decision was made in the former USSR that changed the economic world of the American farmer, probably forever. Grain and oilseed crops in that year were being hurt badly in the USSR by weather problems, and the total crop was expected to fall some 30 to 40 million metric tons short of goals and needs. Historically, when partial crop failure occurred, the USSR liquidated poultry and livestock programs and radically reduced needs for grain and oilseeds. In 1973, they made a different decision. They decided to come into the world market and buy grains. The result of that decision has been documented in earlier chapters. Suffice it to say here that prices for American grains and oilseeds were driven sharply higher, and for the first time since the emergency price support programs were established after World War II, the federal U.S. government had an opportunity to get out of the price guarantee business.

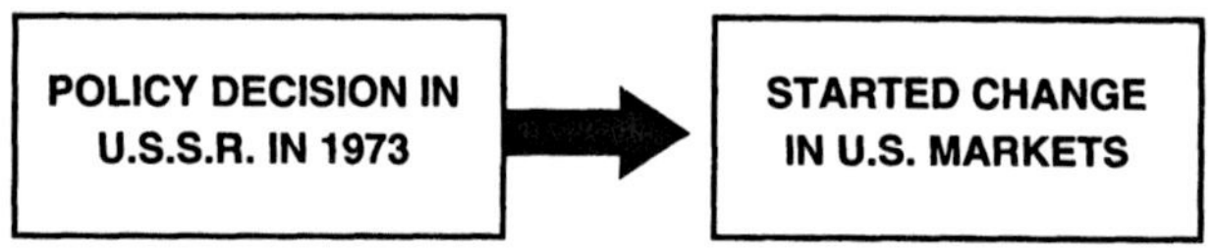

The former Soviet Union fully entered the world market in 1973

FARM POLICY CHANGES

Our domestic farm policy changed dramatically. From production controls and guaranteed prices, we eventually moved to a system of support that characterized the farm policy of 1985 and 1990 and is likely to be continued in some form in the farm policy legislation overall during 1995. In order to participate in the world market and contribute to our overall balance of trade, the U.S. wanted to be able to sell our corn, wheat, soybeans, cotton, and other commodities overseas. A system was established whereby *loan rate or loan price*, which typically was below the world market price, set the price floor for the American producer. For program crops, if price fell below that loan rate, you could simply put the crop in government stocks and if price never recovered, just forfeit the crop as a payment for the loan. But it is important to recognize that, especially in more recent years, those loan rates have been established well below market price levels.

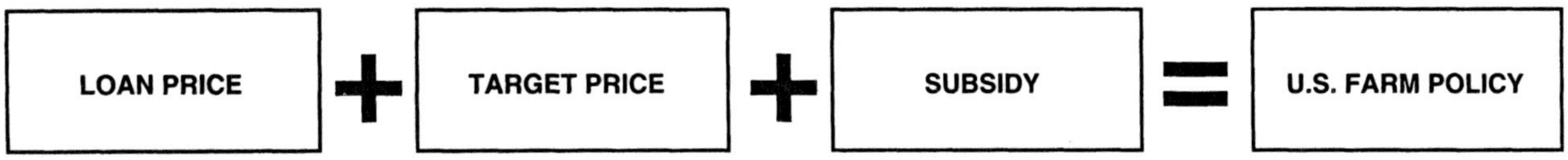

Current U.S. farm policy subsidizes cash up to the target price

The other major dimension of modern farm policy has been the willingness to subsidize back up to a legislatively set price that we know as the *target price*. This means that if prices go below what the legislature in this country had determined as an appropriate price for the farmer, then a subsidy is paid directly to that farmer. This meant the farmer was being helped differently than in earlier decades when market price was guaranteed at some level that was typically well above world market levels. The subsidy came in a different form.

FARM PROGRAM DECISIONS

In current farm policy, the American farmer who is worrying about his marketing plan has to consider whether or not he is going to participate in the programs. Some observers argue that if the farmer is in the program in corn, wheat, or other crops and is therefore guaranteed essentially the equivalent of target price on his specific farm production levels, then there is no need to worry about price risk management. I would disagree with that. Very few farmers have yields established at the farm level sufficiently high to cover all of their exposure. The argument also ignores the tremendous possibility of improving the financial position of the farm in the marketplace. There may be a government subsidy accruing because prices fall below the target price, *but that does not mean that the farmer would not want to take advantage of the opportunity to forward price that crop near the high for the year.* There is also, of course, a number of commodities that are not always in the program. Regarding commodities that for various reasons the yield assigned at the farm level is just too low, the producer may decide not to participate in the program. Then, their marketing expertise is going to be a direct determinant of how well they fare in terms of prices for their commodity.

THE PIK PROGRAM

There are also many of the important agricultural commodities, of course, for which there is no program. This is true for most of the vegetables and fruits and is certainly true for the livestock commodities. Beef cattle is the number one commodity in the United States in terms of receipts at the farm level. Actually, not only is there no price protection in these commodities, but the fact that there are government programs that influence how many acres are planted to corn and to other feedgrains feeds over into the livestock sectors in terms of influencing costs. A very famous example of this policy interference occurred in 1983. After the massive expansion in the 1970's, prompted by the USSR purchases, huge surpluses were starting to build up in corn, wheat, soybeans, and many other storable commodities in the early 1980's. In 1983, the federal government announced its much-discussed **PIK program** (the payment in kind program). This basically allowed farmers to bid their entire farms out of production, and many took advantage. Acreage planted to corn, for example, was reduced over 30 percent compared to 1982 levels--and then the U.S. was hit with a widespread drought.

Corn prices skyrocketed, moving to $4.00 again for the second time in the history of the United States. Producers who were relying on relatively low-priced corn, and livestock, poultry, swine, and dairy programs were hit hard by the price impacts of what turned out to be a disastrous farm policy decision. Thus, we have clear examples that policy considerations can bring volatility to the marketplace for the producer of non-program animals, even many fruits and vegetables.

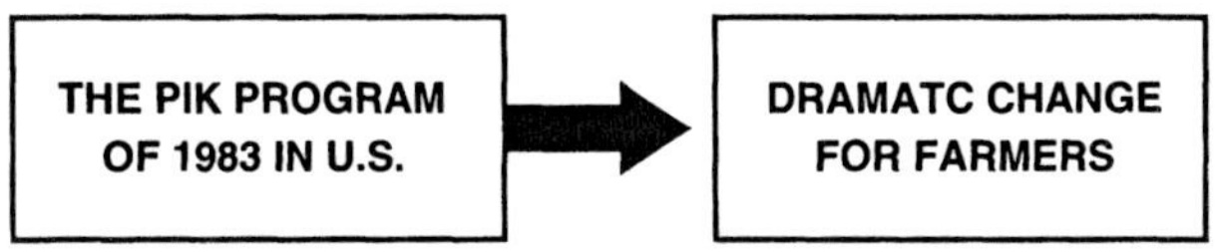

The 1983 policy move proved to be a major disruption

EXPORT ENHANCEMENT

Policy comes in many other forms in terms of influencing the marketplace. In recent years, the **Export Enhancement Program** (EEP) has been a major factor in prices for wheat. The program was established by the U.S. government to allow us to go back into markets where other countries may have traded unfairly to reestablish our position as a seller in those markets. Basically, this means that we sell wheat to countries that need it at a subsidized level. The policy decisions on how many dollars are put into the EEP and what countries those monies are extended to become a very important determinant of wheat prices.

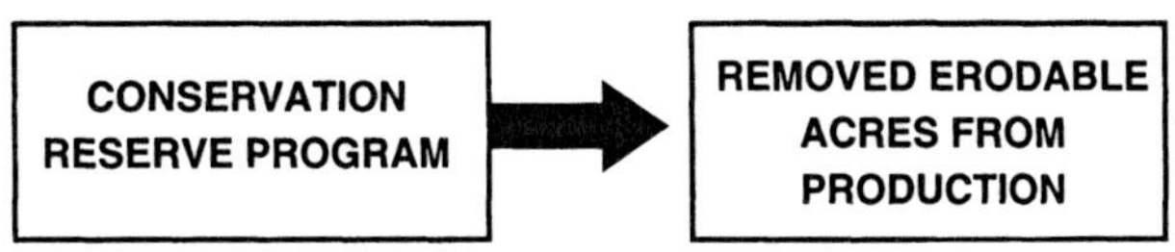

The CRP program took fragile acres out of production

Recent foreign policy programs have influenced prices and the marketplace for program commodities. The long-term Conservation Reserve Program (CRP), started in the mid-1980's, was designed to take land that was easy to erode out of production. As we noted in earlier chapters, many of these acres may have been some of the same 50 million acres that were attracted into production in the 1970's after the USSR purchases drove grain prices sharply higher. That program, by

1994, had taken some 37 million acres out of production. But as discussions were launched with regard to the 1995 Farm Bill, there were no clear assurances that the 10-year contracts, which will start to mature in 1995 and 1996, would be continued in the new farm policy program. In other words a program launched in the name of conservation might be terminated after a 10-year duration, and many of those acres would be expected to come back into production of corn, wheat, and other commodities. Obviously, this uncertainty has implications to total supply and will influence prices received by American farmers.

Policy is, therefore, a source of uncertainty in the marketplace. It is always uncertain that programs such as the CRP program or the EEP program will be extended in the next farm bill. As environmental and conservation considerations become more important to the American public, there is always the possibility that more monies will be put in these programs and less made available to ensure income security at the farm level. All of these things become influences in the marketplace that the individual American farmer has to consider.

Being able to anticipate micro-macro traps is important

AN UNCERTAIN PRICE OUTLOOK MEANS MARKETING SKILLS ARE CRITICAL

What we have, then, are all kinds of reasons to argue that markets and the marketplace will be uncertain and volatile. American farmers thus have a major challenge in front of them. That is why this book is being written. Trying to help the individual producer do a better job of marketing is the main objective.

THE MOST IMPORTANT LESSON

What we found as the chapters unfolded is that prices are much more volatile and that the uncertainty and risk associated with market exposure is growing. Part of this is intrinsic to our industries. The micro-macro trap, for example, that was widely discussed in earlier chapters and was the sole point of discussion in Chapter 3, comes from internal sources. The fact that individuals are price takers in the marketplace becomes makes it critically important develop skills so they can anticipate and spot the coming of those micro-macro traps. *Arguably, that is the single most important lesson that the reader can learn from this book.* Having identified the micro-macro issues and looked at the reasons for price movement over time, whether cyclically or seasonally, we then set out to develop some strategies that can be employed to help the individual producer. Because individuals are price takers and cannot set level of price, they have to focus on controlling the one thing they can control, and that is when they price. But in the process of doing that, we learn that there is a tremendous opportunity to improve the financial well-being and the economic viability of the farm.

POLICY AS A SOURCE OF UNCERTAINTY

Although it was not subject to a great deal of discussion and coverage in earlier chapters, we have indicated in this closing chapter that policy decisions are a major reason for volatility in the marketplace. It may be that the need in the policy arena is not unlike the need guarding against the micro-macro trap. It is important that the individual producer be well read and keep up with the major themes that are involved in the policy process. As we move through 1995, for example, that means being aware that environmental and conservation pressures are likely to continue influencing what is done in farm policy legislation. It may be that over time, some funds will move from supporting and protecting farm income toward financing environmental and conservation programs that are increasingly valued highly by society. *It doesn't mean that any one position here is right or wrong, it just means that the individual producer has to be aware of what is influencing the marketplace in which he or she is going to have to function.*

TRADE POLICIES

In the 1990's we have seen non-farm policy considerations start to have a major influence on the marketplace facing the American farmer. The General Agreement on Tariffs and Trade (GATT) will make a decided difference in terms of our exposure to the world market. In general, the treaty will reduce barriers to trade and encourage flow of goods and services in a world community that is brought closer together everyday by improved communication and transportation technologies. One of the major changes associated with the GATT negotiations is the opening of European common market countries where rather high barriers to import of grains and oilseeds have been long maintained. At the same time, limitations on importing of commodities into the U.S. will go down over time under the provisions of the GATT negotiations. All of this means the playing field is more nearly even, and whatever country can be most competitive in serving the market in a particular corner of the world is likely to get a share of that market. This means, again, exposure to world market considerations for the American farmer.

GATT + **NAFTA** = **INCREASED WORLD TRADE**

Major trade policy exposes U.S. farmers to more of the world market

The more recent negotiation of the North American Free Trade Agreement (NAFTA) is another example of policy and how it will impact the American farmer. This basically opens up a huge trade area made up of Canada, the U.S., and Mexico, and there are possibilities that it will eventually be extended further into Latin America and South America. This means that the American wheat farmer is competing directly with the Canadian wheat farmer for the wheat market in Egypt or wherever the buying market might be. It means that we have the possibility of some processing facilities for grains, oilseeds, and livestock commodities moving to Mexico in a search for lower wage rates. The world will be different after NAFTA, and given that the long-term benefits are almost universally considered to be positive, it will help the farming sector in the U.S. in the long term. But that does not mean that some uncertainties and some difficulties will not surface during the transition period as we move toward full enforcement of the program.

It is going to be a different world and a difficult world. The markets are going to be more challenging, more risky than ever. It behooves the American farmer to step back from this and find a marketing program that fits his or her situation and use it. With that thought in mind, it is time to move to a closing thought on what differentiates the successful marketing program from the unsuccessful one.

A MAJOR REQUIREMENT FOR SUCCESS

In earlier chapters, we talked about some basic tools that can be used to analyze the markets. Fundamental analysis, we found, deals with the basic underlying supply-demand situations. There are some simplistic developments that showed us how to understand why price will move when supply shifts. And we found that over the long run, it may be demand, not supply, that determines the direction of price. Those fundamental considerations are important, and every American farmer will have to understand that basic economic supply-demand framework. It is from the interaction of supply and demand, after all, that we get the prices at which we sell our commodities.

There was also considerable attention paid to technical analysis of the markets. This is especially important in those commodities where futures trade is involved, and that includes most of the major commercial farm commodities in the United States. We found that technical analysis, using some very simple measures, can help us understand the direction of the price trend and the direction in which price is likely to go. Very importantly, technical analysis was presented as an important aid to determining when forward prices or price protection should be established. We need to keep in mind that the individual farmer is a price taker, controlling the "when" of pricing is very important. Technical analysis definitely helps in that process, and there is no reason why fundamental and technical analysis cannot and should not be combined. As a matter of fact, one very basic but solid approach is to use some appreciation of what is going on in fundamental analysis to determine probable direction of price trends, and then to apply technical analysis to help determine when the product should be forward priced.

DISCIPLINE, DISCIPLINE, DISCIPLINE

All of that means little, however, if the individual producer cannot bring a substantial dose of discipline to his or her program. It doesn't do any good to monitor the prices that the marketplace is offering, and be poised to make that critically important forward pricing decision, if the thing that dominates in the final analysis is that inherent insistence to get the top price for the year. If the producer who forward prices his hogs by selling futures at $56 per hundredweight is going to berate himself and feel like he has made a huge mistake when the cash market and the futures market later go to $60 per hundredweight, then carefully constructed marketing plan is likely to be abandoned. You cannot operate the modern farm business as if it were a crap shoot. *You cannot successfully operate the modern farm business if the thing that dominates your thinking is the desire for the high price of the year or the occasional year in which astronomically high prices might be offered. If that is what drives you as a marketer, you are likely to be financially ruined by the bad years while you are waiting for that one spectacular year.* Agriculture in the United

States is a very competitive business. There are no barriers to entry, the micro-macro trap looms, and policy considerations can be a factor in prices. It is very important that the individual entrepreneur learn to step back from all this and start looking for profit margins and start operating the farm business as if it were a business.

Discipline is the key to marketing success

Discipline is the key. Most non-farm business managers would be astounded if they truly understood the level of risk exposure that the farm business tolerates. Most non-farm business managers would be astonished that the farm business manager often operates across an entire lifetime as a cash market speculator. The farmer might come back and say you have to be willing to take some risk and you have to be something of a gambler at heart to be in farming to begin with. Further, you have to be in a position to get those occasional high prices to offset the bad prices that come along. But all of that is short-sighted and wrong. What disciplined marketers will do is take the attractive, profitable price when it is offered by the marketplace by forward pricing (and it really doesn't matter whether it is a cash contract, futures, or options) and then get on with the business of being the most efficient producer they can be. They stay informed, work on understanding the markets, work on sharpening expertise as a marketer, and then take advantage of that. They are perfectly willing, for example, to forego the occasional $60 hog market if nine-tenths of the time they have their hogs forward priced across the years at a significant profit margin.

HAVE A PLAN

If you examine the frequency with which the markets tend to offer at least break-even prices some time during the production period, you will recognize that the opportunities are there in virtually every agricultural commodity for the farmer who is willing to take advantage of them. Thus, perhaps the key requirements for a successful marketing program is wrapped up in the single word "discipline." *If you develop a plan, know what you ought to do, know what you need to do, and then do it, you are likely to be successful and you are likely to be financially independent and economically viable over time.*

SUMMARY

This chapter was, in some respects, a "summary" of the entire book. It is useful here to emphasize a few key points:

- **Farmers face increased uncertainty as we move to a global marketplace.**
- **Policy decisions, domestic farm policy, and trade policy can add to the risk facing American farmers.**
- **The marketplace is changing as our domestic economy restructures.**
- **Farmers increasingly face levels of price risk that most observers would deem unacceptable.**
- **The key to effective marketing and price risk management is discipline--and the potential reward is huge.**

It can be done, and it will help the farmer be viable and competitive.

Appendix

A SELECTIVE BIBLIOGRAPHY

Angell, George, 1990. *Agricultural Options (Trading Puts and Calls in the New Grain and Livestock Futures Markets)*, Windsor Books, Brightwaters, New York.

Chicago Board of Trade, 1989. *Commodity Trading Manual*, Education and Marketing Services Department of the Chicago Board of Trade, Chicago, Illinois.

Chicago Mercantile Exchange, 1990. *Self Study Guide to Hedging with Livestock Futures*, Chicago Mercantile Exchange, Chicago, Illinois.

Chicago Mercantile Exchange, 1989. *An Introduction to Futures and Options*, Chicago Mercantile Exchange, Chicago, Illinois.

Chicago Mercantile Exchange, 1986. *Trading Tactics*, Chicago Mercantile Exchange, Chicago, Illinois.

Cox, John D., and Mark Rubinstein, 1985. *Options Markets*, Prentice Hall, Englewood Cliffs, New Jersey.

Dupont, David B., 1986. "Buy, Sell: Do You Have an Option?" in *Trading Tactics: A Livestock Futures Anthology*, Chicago Mercantile Exchange, Chicago, Illinois.

Eades, Simon, 1992. *Options Hedging and Arbitrage*, Probus Publishing Company, Chicago, Illinois.

Fink, Robert E., and Robert B. Feduniak, 1988. *Futures Trading: Concepts and Strategies*, New York Institute of Finance (Prentice Hall), New York, New York.

Kenyon, David E., 1984. *Farmer's Guide to Trading Agricultural Commoditity Options*, ERS-USDA, Washington, D.C.

Todd Lofton (ed.). *Trading Tactics: A Livestock Futures Anthology*, Chicago Mercantile Exchange, 1986, 153 pages.

Purcell, Wayne D., 1991. *Agricultural Futures and Options*, Macmillan Publishing Company, New York, New York.

Purcell, Wayne D., and Michael A. Hudson, 1985. "The Economic Roles and Implications of Trade in Livestock Futures," in *Futures Markets: Regulatory Issues*, American Enterprise Institute, Washington, D.C.

Purcell, Wayne D., July 1988. *Guide to Price Risk Management for Agricultural Producers and Lenders*, Agricultural Economics, Virginia Tech, Blacksburg, Virginia.

Schwager, Jack D., 1984. *A Complete Guide to Futures Markets: Fundamental Analysis, Technical Analysis, Trading, Spreads, and Options*, John Wiley & Sons, New York, New York.

Glossary of Terms

A

Antitrust Policy
A set of federal statutes that goes back to the 19th century that is designed to ensure competitive pricing and competitive behavior in the U.S. marketplace.

B

Basis Contract
A contract between a seller and a buyer that calls for delivery in the future of some specific commodity at a price level that is tied to the underlying futures contract via a negotiated level of the cash-futures basis.

Basis Risk
The chance that the basis that is employed in looking at the profitability of a production or storage operation and in the hedging or forward pricing strategies will not actually be realized at the close of the program.

Bid-offer Process
The process that describes how futures contracts have traded at the organized futures exchanges as buyers bid a certain price level and sellers or potential sellers offer the futures contracts at a certain price level.

Break-even Costs
The costs that have been incurred in a particular course of economic activity and the cost that is equivalent to the price that would be needed to break-even in that particular activity.

C

Call Option
An option that gives the buyer the right to a long (or "buy") position in the underlying futures contract at a specific price called the strike price.

Cash Contract
A contract that is negotiated in the cash market that ensures the seller of a particular price when he or she delivers a commodity, as described in the contract, to a buyer.

Cash Costs
The variable or "out-of-pocket" costs associated with a particular course of economic activity and exist only when that economic activity is being pursued.

Cash Flow
The in and out movement of cash into and from a business' account with particular emphasis on whether the inflow is sufficient to meet outflow needs.

Cash-Futures Basis
The difference between the cash market and futures market for a particular commodity in a particular geographical area at a particular point in time. Specifically, the cash-futures basis is calculated by subtracting the futures price from the cash price.

Cash Market
The traditional market at which buyers and sellers negotiate a cash price for exchange of a product.

Cash Market Speculator
Someone who is exposed to the movements up and down in the cash market as a producer, a holder of inventory, or a buyer of inputs with no price protection in place.

Closing Basis
The basis at the end of the production or storage period when the cash program and futures trades are being closed.

Commissions
The charge assessed by commodity brokers for trading a futures or options contract for their customers.

Conservation Reserve Program
Referred to as CRP, it is a government program that is designed to pay producers to take fragile and erodible land out of production and put it in a "conserving" use.

Consolidation
The trend toward fewer and larger firms in the processing sector and in the production sector of agriculture; the term is something of a synonym for "concentration."

Consumer Demand
The willingness of consumers to pay varying prices for the quality and quantity of product they are being offered.

Contract High
The highest price recorded during the trading life of a particular futures contract. The terminology "contract high" is synonymous with "life-of-contract high."

Contract Low
The lowest price recorded during the trading life of a particular futures contract. The terminology "contract low" is synonymous with "life-of-contract low."

Cost Ceiling
The maximum price established on an input for an agricultural production or processing program by using call options.

D

Deflated Price
Refers to an inflation-adjusted price where the influence of overall price inflation has been removed from the prices or from the price series.

Demand
A schedule of the quantities that will be taken by buyers at alternative prices.

Demand Shifts
Changes in the entire schedule of quantities taken by buyers at alternative prices when income, preferences,, or prices of related products change.

Domestic Demand
The component of global demand that is U.S. in its origin, specifically in supply-demand balance sheets.

E

Economies of Size
The tendency for costs of an economic activity such as production or processing to go down as the operation gets larger.

Equilibrium Price
The price at which the quantity that prospective sellers bring to the marketplace is equal to the quantity that buyers will remove from the market, a "market-clearing" price.

Expansion Phase
The part of the cycle in livestock production that involves an expanding of the breeding herd.

Export Demand
The demand by international buyers in the global market for agricultural or other commodities.

Export Enhancement Program
A program designed to subsidize shipment of agricultural commodities from the U.S. into selected markets around the world.

F

Farm Policy
A set of legislation that is designed to meet certain policy goals for the farm sector, renewed or reconsidered every five years.

Financial Analysis
Analysis of the financial condition of a firm, often involving debt-to-asset ratios.

Financing
The arranging of money to allow some economic activity to be carried on.

Fixed Costs
Those costs that are independent of the level of output employed and are present whether there is activity or not.

4-Firm Concentration
A measure of the percent of business activity that is conducted by the four largest firms in an economic sector.

Formula Price Contract
A contract calling for future delivery of a particular commodity to a buyer with the price tied to some observable (typically cash) price series.

Forward Price
The price that is being offered by either a cash market buyer or in the futures market for later delivery of a commodity.

Forward Cost
The cost at which input or raw material needs for future time periods can be established.

Futures Contract
Contracts calling for later delivery of a particular commodity that are bought and sold on organized futures exchanges.

Futures Exchange
An organized business activity that serves as a clearing house for buyers and sellers as they trade commodity futures contracts and discover prices for future time periods.

Futures Market
The market that is created and operated on futures exchanges as prices for future delivery are discovered via buying and selling activities.

Futures Price
The price that is being discovered for a particular commodity for a particular delivery period in the future on an organized futures exchange.

G

GATT
General Agreement on Tariffs and Trade, which refers to a worldwide negotiated trade policy.

H

Hedge
The process of taking an equal and opposite position in the futures market vis-a-vis a cash position.

I

Input Costs
Costs associated with inputs or raw materials going into a production or processing operation.

L

Liquidation Phase
That phase in the livestock cycles where breeding stock is being liquidated and sold for slaughter purposes.

Long Hedge
A hedge that involves buying futures contracts or buying call options to protect against rising costs of input.

M

Margin Call
The additional monies that must be sent to keep a futures position in place when the market is going against the hedger and losses are accumulating in the futures account.

Margin Money
The monies that have to be on deposit in an approved and authorized commodity account before a hedger or a speculator can trade commodity futures.

Margin Requirements
The specific levels of monies on a per-contract basis that are set at a minimum level by the exchanges and administered by the various members of the exchanges through their brokerage business.

Marginal Cost
The change in cost associated with another unit of output.

Marginal Revenue
The change in revenue associated with another unit of output.

Market Analysis
Analysis of markets to determine the likely direction of price movement and/or to determine when a buy or sell action should be taken.

Market Indicators
Measures which imply direction of price movement and/or the likely timing of price changes.

Market Power
The ability to influence price or other terms of trade in your favor as a buyer or as a seller.

Marketing Plan
A specific and preferably written plan that delineates how a particular business program intends to manage exposure to price risk.

Micro-Macro Trap
A phenomenon that occurs in a competitive market when the individual is too small to influence price but the combined or macro actions of many small producers drives price up or down.

N

NAFTA
North American Free Trade Agreement, which is a negotiated trade policy involving Canada, the U.S., and Mexico that tends to reduce or eliminate barriers to trade and promote free trade across borders.

Nearby Futures
The closest commodity futures contract in terms of month for which the price is being discovered.

Non-storable Commodities
Commodities such as a fresh fruit or vegetable or livestock in a production program that cannot be stored for any significant period of time without having the quality and/or form of the product change.

O

Opening Basis
The cash-futures basis at the beginning of the production period or at the beginning of a storage period when some course of economic activity and/or futures trades are being considered.

Option Premium
The premium paid to buy an option, either a call or put option, which confers certain rights in the underlying futures market.

Overbought Market
A market that is losing its momentum in the sense that most traders who would be interested in buying the market already have bought or have long positions in futures.

Oversold Market
A market that is losing its momentum to the downside in the sense that most traders who would be interested in selling the market have already sold or have short positions in futures.

P

Payment-in-Kind (PIK)
A program that uses the commodity itself as payment to undergo a particular course of action. The most visible example was the PIK program of 1983 when "payment-in-kind" was employed to encourage farmers to take land out of production.

Price Cycle
The tendency for production of an agricultural commodity to move up and down over time driving prices up and down in an inverse fashion.

Price Discovery
The dynamic process by which buyers and sellers seeking a mutually acceptable transaction price discover the price for a commodity.

Price Floor
The minimum price that is established after appropriate adjustments for basis by buying a put option at a specific strike price.

Price Risk
Essentially "price change risk" in the sense that movement or volatility in prices could be negative for the income and/or revenue position of a firm.

Price Risk Management
Efforts to manage exposure to price risk by making conscious decisions about when price protection will be needed via hedges or options and when it might be best to be a cash market speculator and be exposed to price risk because of attractive income potential.

Price Taker
The terminology applied to a decision maker in a competitive market where each individual is too small to influence price whether in the selling or output market or the buying or input market.

Price Variability
Refers to the tendency for commodity prices to move over time and is essentially synonymous with price volatility.

Pricing Efficiency
The effectiveness of the marketing system in transferring from consumer to producer price incentives to prompt needed changes in quantity and quality of product produced.

Pricing to Value
The ability of the pricing system to price agricultural commodities consistent with their final value at the consumer level.

Production Risk
The riskiness associated with yields, gains, and other dimensions of production as distinguished from price risk.

Profit
The difference between revenues and costs; often total revenue minus total costs.

Put Option
The right to sell or be short in an underlying futures contract at a specific price called a strike price.

R

Realized Price
The price that the hedger nets after completing a cash program and closing out the futures position and completing the round-turn hedge.

Relative Strength Index
A measure of momentum in the marketplace that is useful in indicating overbought and oversold market conditions.

Resistance Plane
A horizontal plane across past price highs that tends to be difficult to surmount if prices rally back toward that level.

Round-turn
The terminology referring to a sell and subsequent buy back or a buy and a subsequent sell in the futures complex.

S

Seasonal Index
A measure of the price for each month relative to the average price for the year in a seasonal price pattern.

Short Hedge
A hedge that involves selling futures contracts or buying put options to protect against declines in prices for a commodity that will be sold at a later time period.

Storable Commodities
Commodities such as corn, soybeans, cotton, or other commodity that can be stored for a significant amount of time without having the quality and/or the form of the product change in a significant way.

Storage Hedge
A hedge that involves placing a cash commodity in storage and forward pricing or hedging it by selling a distant futures contract.

Strike Price
Refers to the specific increments of price at which put and call options are offered for agricultural commodities.

Strong Basis
A cash-futures basis characterized by a relatively high cash price as compared to the underlying futures price.

Supply Function
A schedule of the quantities that will be offered for sale at alternative prices.

Supply Shifts
Movements in the entire schedule of quantities that will be offered for sale at alternative prices because of a change in the cost of an input, change in technology, or some other change that influences the ability to offer product.

Support Plane
A horizontal plane across a past price low that tends to offer buying support if prices drop back toward that plane.

T

Target Price
A specific price in recent farm policy legislation that is designed to be an appropriate price for producers of a particular commodity. The terminology "target price" can also be employed in a price risk management program and refers to a specific price level at which a hedge or some other appropriate action will be taken.

Technical Analysis
As contrasted to fundamental analysis, technical analysis is based on patterns in past prices and carries the assumption that there is informational value in those past price patterns.

Three-Way Agreement
Agreement between farmer, banker, and commodity broker involving details on how a marketing plan is to be conducted.

Trend Line
A line drawn across two lows for an uptrend line or two highs for a downtrend line on a bar chart of price action in a commodity futures contract.

Time Contract
Contract with the primary purpose of specifying the timing of commodity flows and/or the title transfer.

V

Variable Costs
Very similar to cash costs. They are the costs associated with the act of producing, storing, or some other economic activity and are the costs that would be eliminated if the particular economic activity is not pursued.

Vertical Integration
The combining of two or more of the technically related stages of activity between the producer and consumer under a common ownership.

W

Weak Basis
A cash-futures basis characterized by a relatively low cash price relative to the underlying futures price.

Index

A

B

C

D

E

F

G

H

I

L

M

N

O

P

R

S

T

U

V

W

SOURCES OF INFORMATION AND ASSISTANCE

Publication	Publisher
Cattle	USDA National Agricultural Statistics Service
Cattle on Feed	USDA National Agricultural Statistics Service
Cotton & Wool Situation & Outlook	USDA Economic Research Service
Feed Situation & Outlook	USDA Economic Research Service
Hogs and Pigs	USDA National Agricultural Statistics Service
Livestock, Dairy & Poultry Situation and Outlook	USDA Economic Research Service
Oil Crops Situation & Outlook	USDA Economic Research Service
Prospective Plantings	USDA National Agricultural Statistics Service
Wheat Situation & Outlook	USDA Economic Research Service
World Agriculture Supply - Demand Estimates	USDA World Agricultural Outlook Board

To order any of the above USDA publications, call 1-800-999-6779.

USEFUL REFERENCES ON TECHNICAL ANALYSIS

Wayne D. Purcell. *Agricultural Futures and Options: Principles and Strategies*, MacMillan, 1991, 363 pages.

Education Department, Chicago Board of Trade. *Commodity Trading Manual*, CBOT, Chicago, 1989, 401 pages.

Todd Lofton (ed.). *Trading Tactics: A Livestock Futures Anthology*, Chicago Mercantile Exchange, 1986, 153 pages.

INFORMATION SOURCES

Chicago Mercantile Exchange
30 South Wacker
Chicago, IL 60606
Marketing Department
(312) 930-8213
Education Department
(312) 930-3198

Marketing and Education
Chicago Board of Trade
141 West Jackson Blvd.
Chicago, IL 60604
(312) 435-7217

MARKETING AGRICULTURAL COMMODITIES

This book is written for the current and aspiring American farmer. The concept of marketing can be big and complex. However, the way Dr. Wayne Purcell, nationally known authority on agricultural commodities, has written his book is simple and direct.

The farmer is the first link in the chain of events that becomes a marketing system. The final link is the consumer. An individual farmer has tremendous potential to improve his or her position by becoming better in marketing (more so than by still further improvements in production). This book is dedicated toward providing the basics and the applications that the individual will need.

The focus here is on the marketing decisions of the farm producer:

- **What does it take for that producer to make effective marketing decisions?**
- **How important is it?**
- **Is this the way farmers can differentiate themselves from neighbors and make a decent income when others are struggling?**
- **How much potential is there if the farmer as a decision- maker starts to get as good in making wise marketing decisions as in making smart production decisions?**

As we become more nearly a global market, it is going to be increasingly important that the individual farmer, and farmers in general, become better marketers than they have ever been before. The potential is huge. The opportunities are before us. The need is to recognize that potential and those oppurtunities and to go after them. This book is dedicated to helping you get there.

Dr. Purcell has received eleven teaching awards since 1968, has authored several other texts on the subject, and has written over 200 additional publications and articles. He is a Professor of Agricultural and Applied Economics at Virginia Polytechnic Institute in Blacksburg, Virginia. Wayne also serves as the Director for the Research Institute on Livestock Pricing.

JOHN DEERE PUBLISHING
The all-encompassing source

Deere & Company

FBM14101NC
ISBN 0-86691-217-7
Litho in U.S.A. (J-00)